S0-DZC-014

Arthur Darack and the Staff of Consumer Group, Inc.

SMALL ENGINE MAINTENANCE AND REPAIR FOR OUTDOOR POWER EQUIPMENT

Prentice-Hall, Inc., Englewood Cliffs, New Jersey 07632

Library of Congress Cataloging in Publication Data

Darack, Arthur.
Small engine maintenance and repair for outdoor power equipment.

Includes index.
1. Internal combustion engines, Spark ignition—Maintenance and repair. I. Consumer Group. II. Title.
TJ790.D37 1984 621.43'4 83-24704
ISBN 0-13-813148-1
ISBN 0-13-813130-9 (A Reward book : pbk.)

All illustrations are reprinted courtesy of Briggs and Stratton Corp., Kohler Co., and Tecumseh Products Company, 900 North Street, Grafton, Wisconsin.

Printed in the United States of America.

10 9 8 7 6 5 4 3 2 1

ISBN 0-13-813148-1

ISBN 0-13-813130-9 {A REWARD BOOK : PBK.}

Editorial/production supervision by Peter Jordan
Cover design by Hal Siegel
Cover illustration by Peter Bono
Manufacturing buyer: Pat Mahoney

This book is available at a special discount when ordered in bulk quantities. Contact Prentice-Hall, Inc., General Publishing Division, Special Sales, Englewood Cliffs, N.J. 07632.

Prentice-Hall International, Inc., *London*
Prentice-Hall of Australia Pty. Limited, *Sydney*
Prentice-Hall Canada Inc., *Toronto*
Prentice-Hall of India Private Limited, *New Delhi*
Prentice-Hall of Japan, Inc., *Tokyo*
Prentice-Hall of Southeast Asia Pte. Ltd., *Singapore*
Whitehall Books Limited, *Wellington, New Zealand*
Editora Prentice-Hall do Brasil Ltda., *Rio de Janeiro*

CONTENTS

INTRODUCTION

Small gasoline engines power lawn mowers, riding mowers, snowblowers, and other devices that move around. Stationary tools and equipment primarily use electric motors. There are exceptions: small mowers may have electric motors and gasoline engines may power a few more-or-less stationary devices such as electricity generators and air compressors.

The (amateur) fixer can maintain and repair any or all of these devices with a few tools, some determination, the information in this book, and a few incentives. As to the latter, a call to any shop for prices on the repair of a lawn mower engine should galvanize even a sloth.

When your lawn mower won't start, it isn't exactly a federal case (they're not subject to recall—yet), as it might be when your car misbehaves that way. You can always postpone cutting the grass while you ponder the situation. If you decide to do something about it on your own, however, you face problems similar to those you might encounter in car engines. As with car engines, you need to learn symptoms and procedures. A one- or two-cylinder engine may have fewer components, but they work on the same basic principles as the larger engines.

The first principle is that the small engine needs the right fuel mix, an ignition spark at precisely the right moment, and an engine with compression (unworn rings, cylinder walls, and valves) that allows firing to occur.

If you run a lawn mower, you also keep a can of gasoline. The first test you can make consists of pouring a bit of gas into the throat of the carburetor—if you can find it. Then try starting the engine. If it doesn't start with the extra gas, you almost certainly have electrical problems. To determine if that is the case, pull off the spark plug cable located in the center of the engine. Then, wearing a heavy glove wrapped in a towel, hold the cable about a quarter of an inch from the plug and crank the starter. If no spark occurs, or if the spark is barely discernible, you need new points and a new condenser, or a new magneto or an adjustment on the old one.

If you get adequate spark and fuel flow and yet cannot start the engine, the problem is internal. However, let's concentrate on externals for the time being.

The engine you face is probably a Briggs & Stratton and is less likely to be a Clinton, Tecumseh, or another make. That's because Briggs & Stratton engines have the kind of domination in lawn mower engines that General Motors has (or had, as the case may be) in cars. However, there are not that many differences among the various makes. We will note, as we go along, the emphasis one finds from one company to another. (There is also, inevitably, a Japanese contingent, beginning with Honda.)

TOOLS

Before you commence tinkering with an engine, you must own a few tools. You need a small socket wrench set, which you may buy for a few dollars, but buy only good quality tools. Many stores feature socket sets as "loss leaders." Buy the standard sizes, since U.S. engines obstinately continue to avoid metric. You will also need such extra but versatile and common tools as screwdrivers, vise grip pliers, a small hammer, an adjustable open end ("crescent") wrench, and some other common wrenches. Many people buy tools on impulse, so you may already own most or all of the above. You may own more. A few special tools are sometimes prescribed for small engine repairs, including something to force off the flywheel. That's an inexpensive hub puller that you may wish to buy at a lawn mower service center that also sells parts and tools. As we will see, you can get along without such a device. It depends on how fastidious you are; flywheels can be pounded off without damage if you know how and where to pound, but many people consider such an approach barbaric—about like playing Mozart on a kazoo.

All these tools and others you may require—for example, an electric drill—can be used for many jobs unrelated to lawn mowers. Thus, you need feel no guilt about buying tools for a particular job that doesn't pan out; there will be others.

MAINTENANCE

One usually begins an account of engine maintenance by stressing cleaning and oiling operations. Cleaning consists mostly of brushing away grass and leaves, especially around the cooling fins of the flywheel, but engines benefit from cleanliness almost as much as the old adage recommends it for people.

Oil should be changed only when it is warm, otherwise the sludge that you must remove remains behind and the fresh oil is largely wasted. Ideally, oil should be changed at the end of the mowing season, so that fresh oil will be in the engine sump during the winter, reducing acid formations. For the superconscientious, an oil change halfway through the mowing season would not be unreasonable.

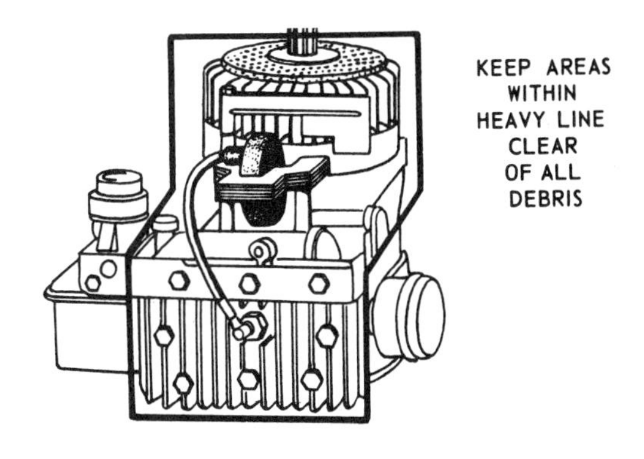

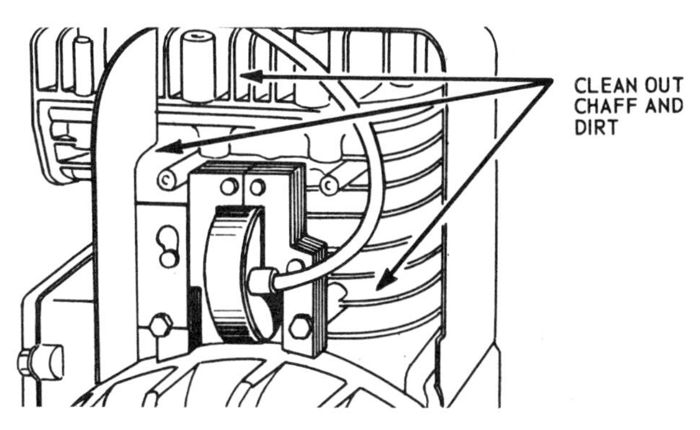

FIGURE 1. A vertical and a horizontal crankshaft. Brush debris away from these and other parts of small engines.

To drain engine oil, prop the mower up on a box or stand so as to expose the oil drain plug underneath the deck, but not so as to spill gas from the carburetor. The plug is usually squared, requiring a crescent or open end wrench for removal. Drain the oil into a pan and examine the cutting blade. If it is damaged, it should be replaced. Above all, check out its retaining nuts and/or bolts for proper tightness. You may want to take the blade off and sharpen it. The retaining nuts or bolts may be rusted or tightened so much that unusual force may be needed to remove the blade. You probably will need to use a socket wrench. You may also need to use a long rod extension bar on the socket for extra leverage, as well as some penetrating oil. To avoid the remote possibility of the engine starting, pull off the spark plug cable and tie it away from the plug. To anyone used to the army mule balkiness of small engine starting, such a safety precaution may only arouse derisive disbelief, but people have been injured severely in just such a situation.

The refill cap is on top of the deck, alongside the engine. Usually plastic, it can be turned off by hand, or with a small screwdriver to start it. After replacing the drain plug, fill up the sump until oil reaches the indicated level. If there is no mark, fill it almost to the top.

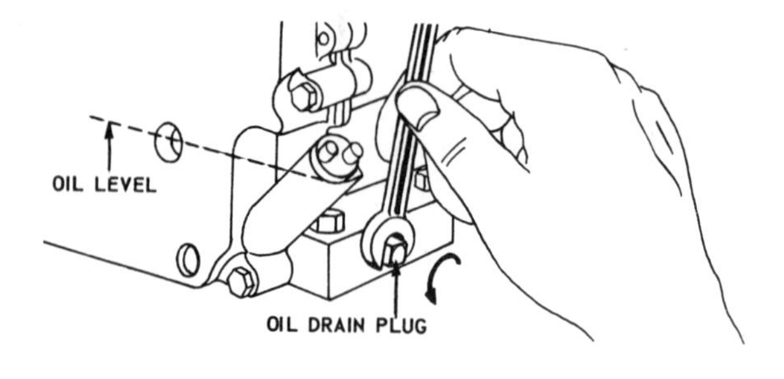

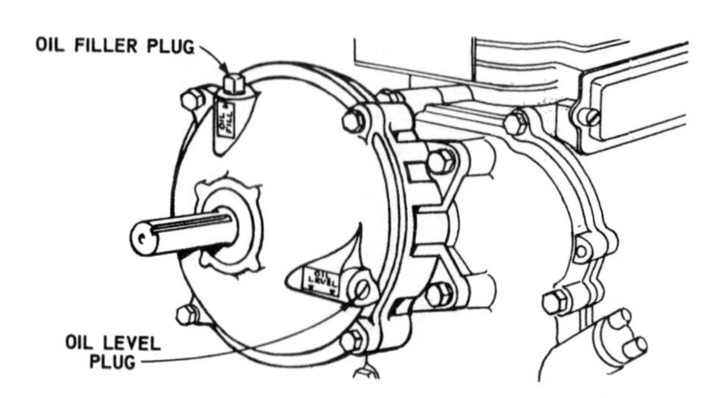

FIGURE 2. Oil filler and drain plugs. Drains are found at various places, always at the bottom of engines; filler plugs higher up.

It is impossible to describe the range of blades one finds on lawn mowers. Mower blade design brings out great ingenuity for so humdrum a task. The latest wrinkle is the plastic blade consisting of "fingers" that cut the grass but not feet, shoes or other human attachments. However, the safety of this design exceeds its utility and its future is clouded. Whether people who put their feet under a whirling mower blade actually exist is unknown, but if they do it isn't likely that they would want a device such as the plastic blade. On the other hand, there are eccentric blade designs that one does need to look at with safety in mind. The large steel wheels with small blade appendages that once were all the rage should be examined very carefully. You deal with three or four potential guided missiles on these things, and if you own such a mower you would do well to test the blade wheels for tightness at least once a summer. One thing you must do, if you are interested in safety, is to track down any unusual noise or vibration. Vibration suggests that something is coming loose. Noise indicates that something *is* loose. It may be nothing to worry about, but one doesn't ignore symptoms of this kind in any motorized device.

The most common cause of instability and noise arises from engine bolts coming unhinged. These bolts, anchoring the engine to the deck or the engine shield to the engine or deck, are loosened by engine vibration. They should be tightened every summer.

Wheels should be examined for loose axle nuts and bolts. They should also be greased or oiled periodically. Other regular maintenance chores on a lawn mower are limited to cleaning and oiling the sponge atop the carburetor housing, if your engine has such a system, and every two or three seasons lubricating the accelerator mechanism, including the controls and the cable. Graphite is the usual lubricant for cables, but if the cable is worn excessively it is best to replace it with a new one. They are inexpensive and easily installed.

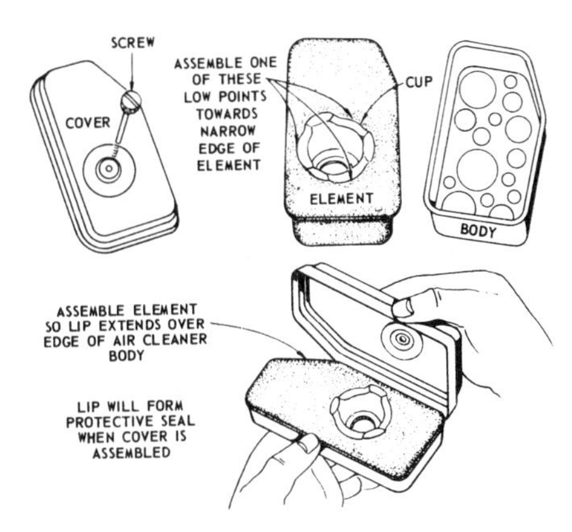

FIGURE 3. An oil-foam air cleaner.

Some small engine manufacturers use dry element carburetor air cleaners, as in car engines. Once a season, these should be taken out and tapped on the pavement to loosen the dirt. If dirt is noticeably present it is well to buy a new one.

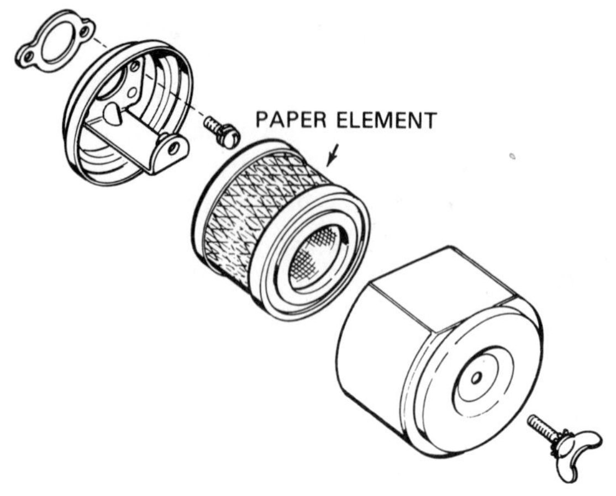

FIGURE 4. A paper element air cleaner.

ELECTRICAL PROBLEMS

Ignition service on a small engine isn't much easier than on larger engines. It's just different. Hence, the inflated charges you may pay when you have a professional tuneup on your mower engine.

The first service should attend to the spark plug. For that you need a spark plug socket. That's a separate tool from the sockets in your regular tool set, and you also need a gapper.

Remove the spark plug cable by twisting it gently a half-turn, and then pulling it off. (Twisting eases the disconnection.) Remove the plug, using the special socket. If the spark plug electrodes are black and oily the engine needs work—later. With a gapping tool (or feeler gauge, as it is called), check the distance between the two electrodes of the plug. Set the gap by bending the outer electrode in or out, according to specifications on your engine, which should be listed on an engine decal. It could be anything from .018 inch to .030 inch.

Clean off the electrodes with sandpaper, emery cloth, or a thin file—anything that will scrape clean all the blackened, corroded surfaces without damaging the electrodes.

If the electrodes show any sign of wear, or the porcelain is cracked, the plug must be replaced. The electrodes should be in good shape; that is, there should be no worn or mangled spots and cleaning should make them look like new. In any case, it is well to replace plugs every other season or more frequently if the mower is used often, say more than once a week. See Figures 6 and 7 for spark plug checking procedures.

Why, you may ask, should any of this be necessary? Spark plugs create combustion. That, in turn, burns them up. It's a slow process, just as dirt and engine vibration change control settings and cause wear throughout the engine. Anything that moves has friction wear. In a small engine there aren't many parts that are exempt, beginning with the hand controls.

After spark plugs, it is well to continue the tuneup of the balance of the electrical system.

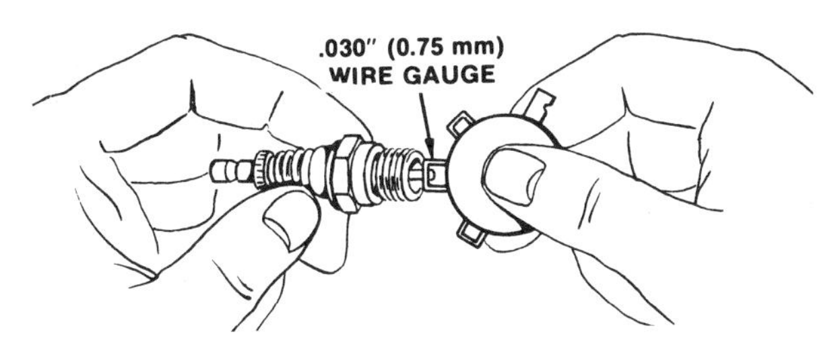

FIGURE 5. Checking and adjusting spark plug gap.

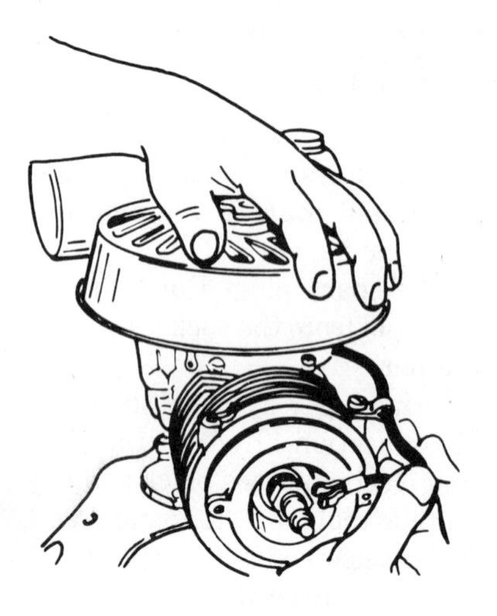

FIGURE 6. To check for spark, hold the cable off the plug and turn the flywheel.

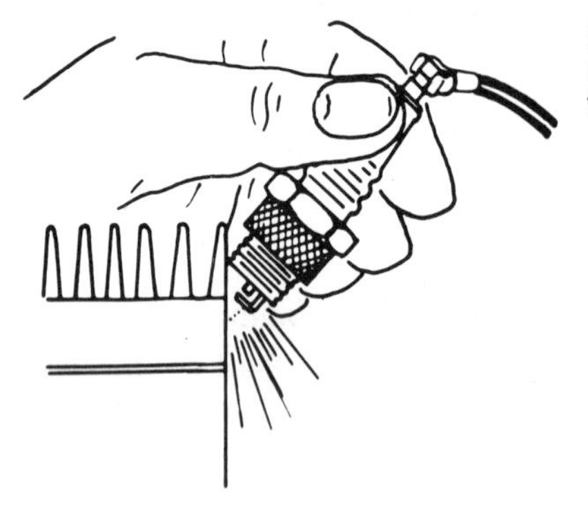

FIGURE 7. To check the plug for firing, remove it, hold it against the engine, and turn the flywheel.

The Briggs & Stratton one-cylinder engine with a rope, spring, or windup starter is the most common power source. It has great durability and ease of repair. Other engines may surpass it in this or that refinement or detail, but none can beat it as a price/utility combination. You can still buy a mower with one of these engines for around $100, give or take a few, whereas you can pay two or three times that much for just the engine if you want a few extra gimmicks or need extra power. However, it should be noted that Clinton engines rival the Briggs & Stratton in price and durability, without having its market clout. Perhaps that is due to their design—a two-cycle stroke instead of four. There is nothing inferior about a two-cycle engine for small engine tasks, but the idea of mixing oil and gas apparently turns a lot of people away, as if it were mixing oil and water.

An electrical system in a single-cylinder engine such as the Briggs & Stratton includes the spark plug, magneto, which produces the spark for the plug, and the points and condenser. Of these elements, only the magneto is not supposed to require regular replacement. In fact, magnetos do indeed last for many years, given normal luck and use, though they may need adjustment of their gaps on occasion. However, the points, condenser, and spark plug are wearing components, just as they are in auto engines where they get replaced fairly often.

One judges the need for points and condenser replacement by eliminating carburetion and the magneto. Also, it is a question of use—how long has it been since the points and condenser were replaced? If more than two seasons, it's time. In additon, hard starting almost certainly is caused by ignition rather than carburetion, though it can be a combination of the two. That's because there is far more wear on electrical components than on the fuel system in a small engine.

Before setting out to replace points and condenser in a Briggs & Stratton, Clinton, Tecumseh, or any other engine, make sure the new set you buy is the correct one for your mower. You buy any part by engine serial number and part number if the part is so identified. It helps, of course, to take the old one along, and once you learn how to extract it, that is the best policy.

To get at the points and condenser requires peeling away several layers of components, including any shroud or covering—usually some plastic or other screwed-on outer skin. Covering aside, the engine begins with the starter and the housing above the flywheel. A screen to keep out debris is often above the flywheel. It is held in place with several small screws. A small socket is best used on these screws. Though they are often made to accept a screwdriver, you do well to use the more protective socket. Professional mechanics may overtighten these things and a screwdriver will ruin them.

The major removal consists of unbolting the housing over the flywheel that contains the starting mechanism—rope or cranking wind-up. That housing is bolted on the sides of the engine and at the front. In fact, you cannot remove the screen until you get the starter housing off. To unbolt the housing, use a socket wrench on the four or five bolts that retain it. You will also have to unbolt the handlebar control bracket at the starter housing end. Notice how the fastener works here, because it is both fastener and adjuster. It lets in and takes out "play" in the control cable, and you want to put it back about where it was after you are finished.

With the housing and screen removed, you now face the starter windup clutch and, below it, the flywheel. The starter clutch screws on over the top of the crankshaft that you have just exposed. To get it off requires force, but the clutch is soft metal, especially at the three notches provided for removal. Special tools can be bought to do the job, but a small block of wood and a hammer will do as well. The wood block must be sturdy enough to accept stout blows. A 2 × 4 block, 10 or 12 inches long, will work. Put one end against each of the three notches in turn, and hit the other end of the wood block, with the force being administered counterclockwise. That will loosen the clutch; now you can turn it off by hand.

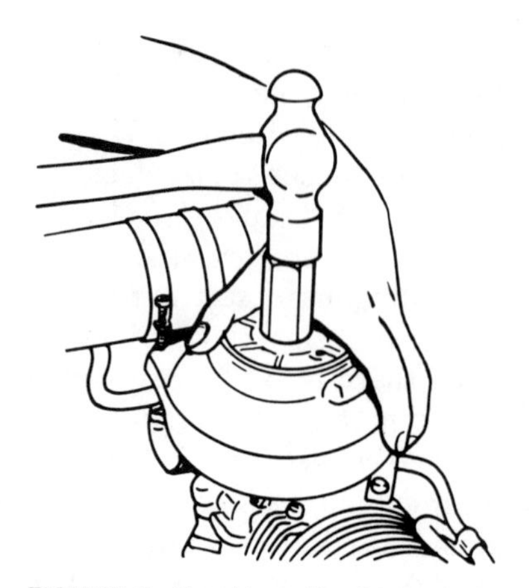

FIGURE 8. Knocking off a flywheel.

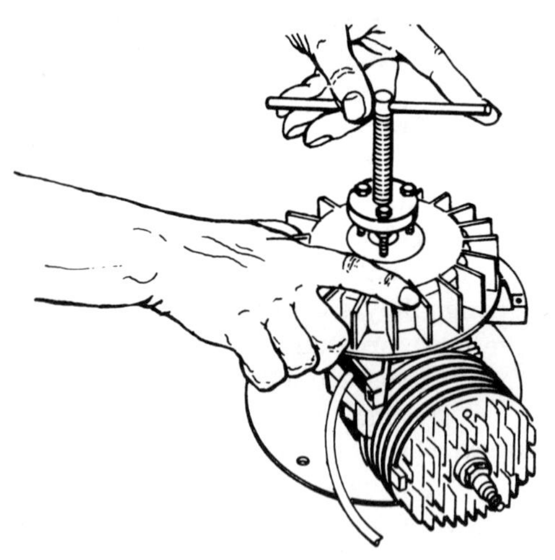

FIGURE 9. A Tecumseh flywheel puller.

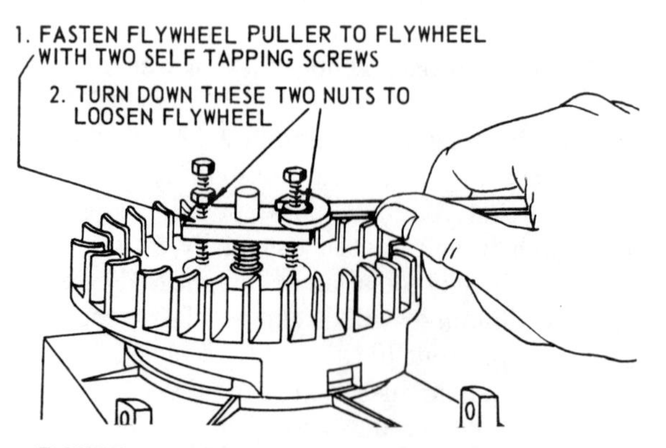

FIGURE 10. A Briggs & Stratton flywheel puller.

Next comes the flywheel, held in place in a press fit with a slotted bar to prevent circular movement. There are two ways to remove the flywheel: (1) buy an inexpensive wheel puller, or (2) use a hammer. If you adopt (2), grasp the flywheel below and pull up on it forcefully enough to raise the mower off the ground, then sharply hit the crankshaft end. It is best to use a lead hammer for this operation, but any other hammer will do. If the flywheel doesn't loosen after one or two blows, either you lack determination or the press fit is too tight. In either case, buy the puller, which costs only several dollars. The reason I don't demand that you buy the puller is because the hammer does work and it doesn't damage the crankshaft. If you buy special tools for everything you do on a lawn mower you risk going into the lawn mower repair business.

With the flywheel removed, notice the slotted key inside the wheel. Usually, when you buy a point-condenser set it also contains a new key, but the old one is perfectly reusable. The key must be replaced exactly as you find it—in the slot.

There is one other method of flywheel attachment and removal, namely a nut on top that you turn clockwise—the opposite of the usual direction—to remove. Once the nut is removed you still have the press fit to contend with. The best, safest tactic is to use two nuts on the top of the post. Loosen the original retaining nut and unscrew it to a point about a quarter-inch below the top of the crankshaft. Then put another nut over that one, finger tight, and hit *that* one while holding the flywheel as prescribed above.

In either flywheel removal operation you must avoid damaging the screw thread on the top of the crankshaft. That means avoiding the thread itself and avoiding excessive pounding on the top, which could cause the bolt to spread and make it difficult to screw the clutch back over the top of the crankshaft.

I mention these possible mistakes not to frighten you away but to note the basic fragility of engines. They may be constructed of various metals, but many of the metals are soft, and even ordinary use can cause a lot of stress and wear.

Below the flywheel, which should now be in your hand, is an aluminum housing that contains the points and condenser. Cables coming out of the housing should be inspected for wear. Insulation wears off from rubbing against the housing. If this has happened, you may have to replace the cables. You can try using electrical tape to repair any worn spots, but the chances are that you'll need new cables. They are soldered at the magneto end, and probably also at the points-condenser end of the line. If the insulation is worn through, that, in itself, will prevent an engine from starting.

Aluminum housing fasteners consist of a couple of small screws that come off with a small socket wrench. Remove the cover, noting the

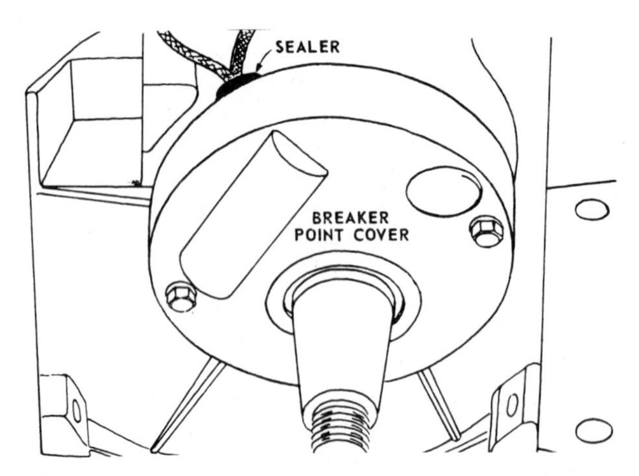

FIGURE 11. Point housing cover on a Briggs & Stratton engine.

tightness of its fit. If there's oil on the floor of the housing, it means that there is a worn point plunger, and/or worn rings. Loss of compression, caused by worn rings, will interfere with starting. While a worn point plunger won't have that effect, it should be replaced anyway, each time you replace the point set or whenever you find oil inside this housing. Usually the point kit will contain the plunger. It pushes open the points and closes them, getting its force from the engine's turning camshaft.

Now that you've exposed the points, turn the crankshaft around—it's perfectly safe to push on the blade below, since the ignition system can't work with the flywheel off—but as an additional precaution you should also remove the spark plug cable. Notice how the points open and close as you turn the engine crankshaft. That opening or gap is important; it must be just right to coax the spark across it and then to fire the plug and the air/fuel mix that is drawn into the combustion chamber by vacuum pressure. The gap will be about .020; your mower instruction booklet will tell you exactly what it should be, but if you

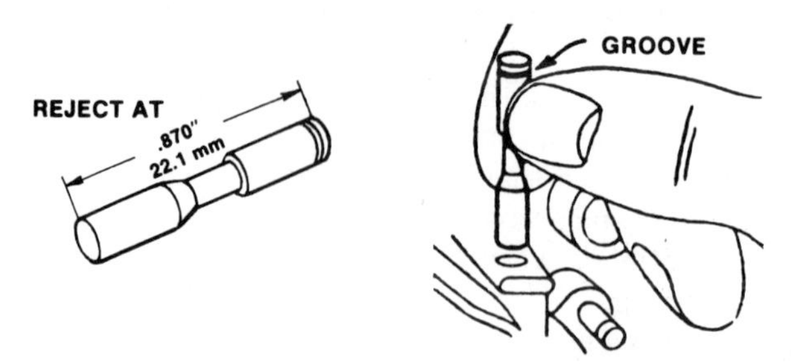

FIGURE 12. Briggs & Stratton distributor plunger.

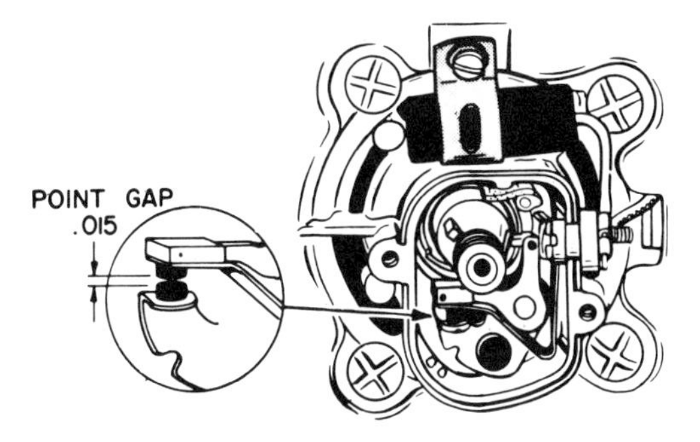

FIGURE 13. A two-cycle Tecumseh point setting.

don't have such information the .020 number will do. When the proper feeler gauge touches both point surfaces, the adjustment is right.

If the points show little or no wear, don't replace them unless the job hasn't been done for over two seasons. Point wear is not the only source of trouble; the condenser can become defective over time, though that isn't common.

In a Briggs & Stratton engine the points and condenser are held in place by a clamp and a bracket, both of which unscrew best when you use a socket on them. At the top of the condenser that holds an electrical lead is a tough little tension spring. The only way you can get the spring down far enough to allow you to remove and replace the electrical lead is to use a plastic tool (a "depressor tool") that should be enclosed in the point-condenser kit. The plastic tool, which is nothing more than a few cents' worth of plastic in the form of a cap that fits over the condenser half of the point set, has only that one task in life. It does not self-destruct, as do those insects that produce a new generation as their sole task and then gracefully leave this vale of tears.

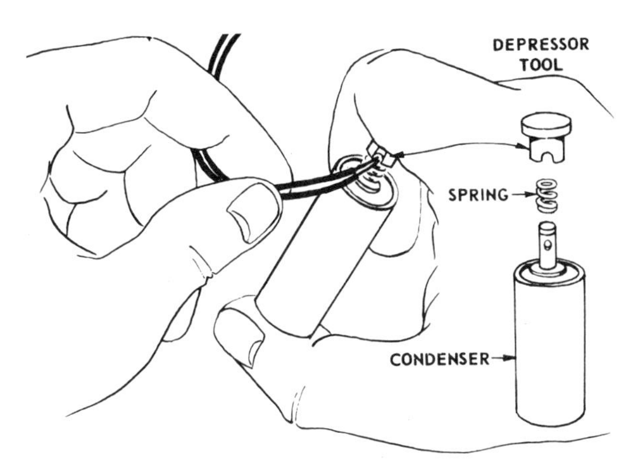

FIGURE 14. Briggs & Stratton condenser tool.

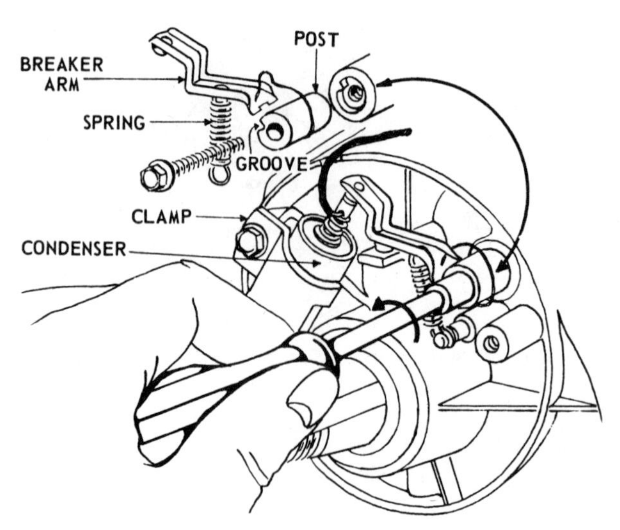

FIGURE 15. Removing point post screw on Briggs & Stratton point set.

Notice how the cylindrical condenser and the praying mantis-shaped movable points are fastened. A tricky maneuver is needed to put in the new set if you don't record in your mind's eye the way you got the old set out. The point bracket that moves with the turning of the engine is fastened with a spring that locks the bracket into a groove. A small bolt holds the post, which you can remove if you find it necessary to install new points.

Flywheels on Clinton engines come off more or less exactly as described above. Snap off the cover of the point-condenser housing (instead of unbolting it), and you have exposed the points and condenser. There is a gasket to be examined on the point cover housing. If it is damaged, it, too, must be replaced.

Clinton engine points are pushed open by the movement of cams that fit over the crankshaft. Some Clinton engines have cams that are part of the crankshaft—essentially like the Briggs & Stratton engine. The manufacturer urges that you notice whether or not the point surfaces come together fully. If they don't, adjust the stationary rather than the movable point. (That's always true.)

One aspect of the Clinton engine differs from other engines: flywheel and crankshaft are tapered to fit each other. If the fitted surfaces are brightly marked instead of being dull and uniformly shaded, the fitting is wrong. To correct the fitting, make sure the key is installed properly and the nut on top of the crankshaft is tight.

When buying point sets it is best to buy original equipment makes. You may save a dollar or so at bargain stores but you may not get the best quality points, in which case you do not save.

In dealing with small engine ignition problems you may also encounter electronic ignition systems. The control unit is supposed to last forever, so you must first suspect the magneto, spark plug, and wiring when no spark can be produced at the plug (by hand-turning the crankshaft and holding the cable a quarter-inch from the plug post as illustrated in Figure 6).

Let us look now at a Tecumseh small engine ignition system.

For purposes of explanation, notice the component parts of the Tecumseh ignition system as they function. The stationary point is not atop the condenser, as in the Briggs & Stratton engine, but separate from it, as in auto distributors. The magneto generates spark (electrical impulses) by means of a stationary assembly (the stator) mounted on the engine that works in conjunction with magnets fixed in the rotating flywheel. In effect, it acts like the rotor of an electric motor or generator. Flywheel magnets turn the magneto's laminated legs. This sets up magnetic force fields that, in turn, generate electricity. These force fields fight each other in reversals (due to opposing poles in the magnets), causing large voltage buildups in the secondary coil of the magneto. Secondary voltages lead to the spark plug (with the other end of the secondary coil grounded to the stationary magneto body). The voltage leaps across the spark plug electrodes on its way to the ground. As it does so, it fires the fuel/air mix in the compression chamber.

Points and condenser function in these processes at the beginning of the spark generation sequence and at the end. The points are switches

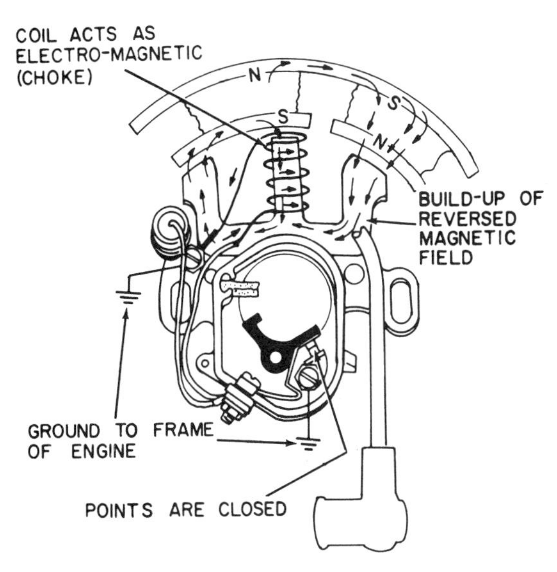

FIGURE 16. Electrical sequence in a two-cycle Tecumseh engine showing points closed.

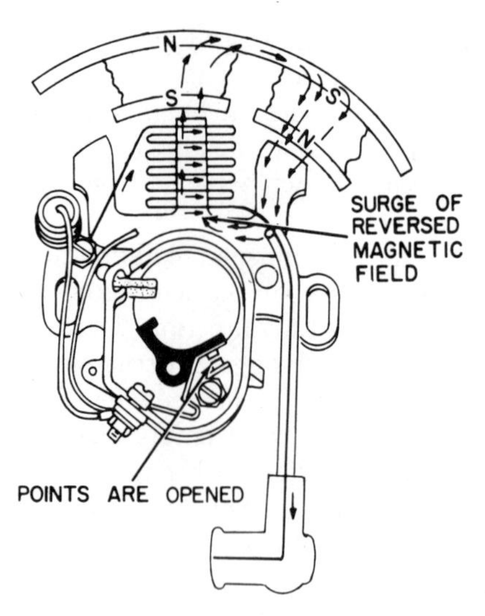

FIGURE 17. Electrical sequence with points open.

that oppose and permit the flow of current. The condenser is an electrical shock absorber that prevents arcing (short-circuiting) of the points when they open. Defective condensers ruin points quickly and make engine operation impossible.

Any magneto has a coil that consists of two windings—a primary made of only a few turns (150 or so) of heavy wire wrapped around center metal laminations, and a secondary winding of very fine wire with many turns (10,000 or so) wrapped around the primary windings. These coils are sealed hermetically inside a plastic casing. One lead of the primary winding connects to the moving contact point; the other is grounded to the stator body—the stationary part of the magneto. The secondary winding connects to the spark plug via the thick spark plug cable and terminal, while the other lead is also grounded to the stator body.

Testing and disassembly of the two-cycle Tecumseh engine differs a bit from other small engines, including the Briggs & Stratton we've talked about. To test hold the spark plug cable with an insulated pliers about ¼ inch to ⅛ inch from the metal part of the spark plug. Crank the engine with the usual procedure. If a bright, hot spark jumps from the cable to the metal body of the plug, the magneto is working. Now remove the spark plug and connect the cable to it. Using an insulated pliers, hold the plug against the body of the engine and crank as usual. If a spark flashes brightly across the plug gap, the whole ignition system is working properly. If the engine doesn't start, the problem is in the fuel system or inside the engine.

Examine the spark plug electrodes. If the electrodes are burned or have a buildup of heavy carbon, or are full of oil, the plug must either be cleaned thoroughly and gapped or replaced, but leaks are the more likely culprit. If the spark plugs have heavy carbon deposits, it means the carburetor setting is too rich—too much gas, too little air. It could also mean the choke valve isn't opened fully during normal engine operation, but it could also be caused by the wrong gasoline, a clogged exhaust or too much oil in the oil/gas mixture of the two-cycle engine formula.

For disassembly to get at points and condenser, if the electrical tests indicate the need for an engine tuneup, remove the engine shroud and starter. The flywheel nut may have either a right- or left-hand thread; examine it carefully. Engines such as these Tecumseh two-cycles can turn in either direction, depending on their use (lawn mowers, generators, etc.). Direction of the thread controls the way you

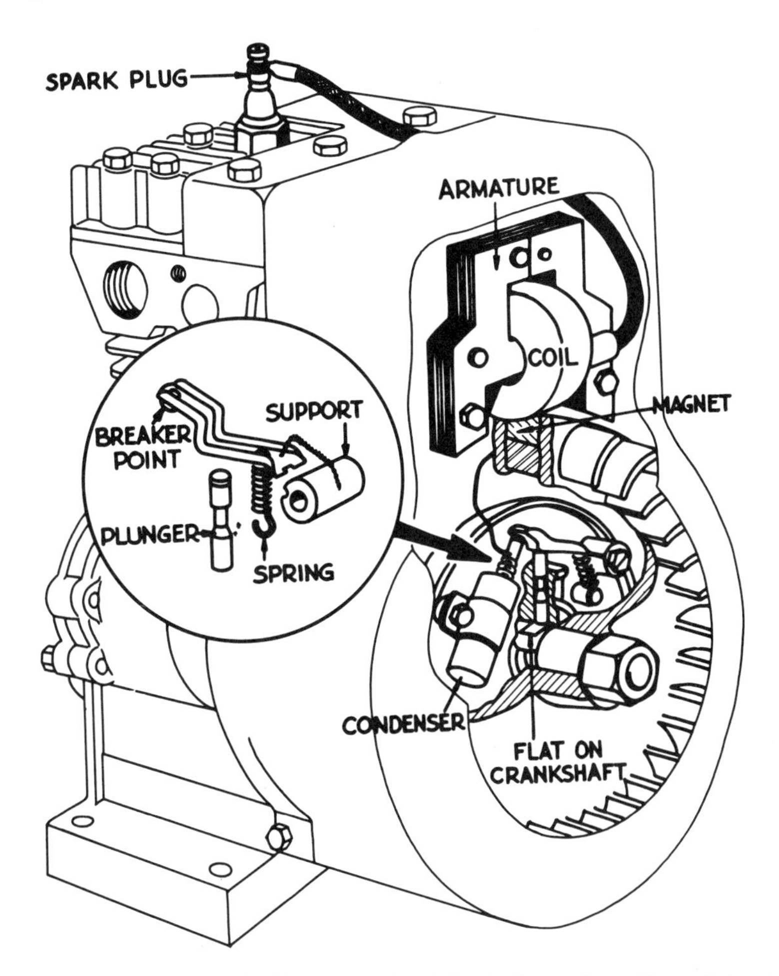

FIGURE 18. Drawing of the ignition system in relation to the engine, with inset of point components.

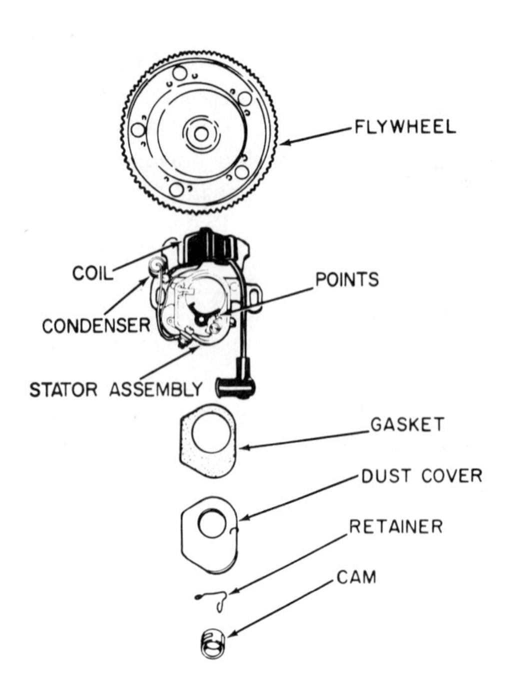

FIGURE 19. Component parts of a Tecumseh two-cycle engine ignition system.

loosen the nut. Once the nut is loosened and removed, the flywheel presents a more pressing problem. It is a press fit and can be removed either by pounding with a special knock-off tool, or with a small flywheel puller if the flywheel has the right configuration of cored holes and the puller has the right kind of self-tapping screws to fit. You can also use the same technique I described for the Briggs & Stratton engine with the nut on top of the crankshaft—put a second nut over that one to protect the threads and absorb the force of the hammer. Then bang it a couple of times, pulling up on the flywheel with the hand that isn't using the hammer. Some Tecumseh models do not allow any banging, either with or without the knock-off tool. These flywheels must be pulled off. Instructions will accompany such engines. Incidentally, don't be afraid to use a blowtorch on these or other flywheels that won't budge. Just don't torch any coils or plastic. Heat the area closest to the crankshaft, because the aluminum alloy of the flywheel expands more rapidly than the steel of the crankshaft under heat, and the seal will be broken between the two components.

To replace points and condenser and set the gap correctly in the new points, remove the nuts from the leads on the movable point spring and remove the point. Next, unscrew the stationary point. Place the new stationary point on the plate but don't tighten the screw just yet. Next, place the new movable point on its post. Adjust the point gap at the high side of the cam, turning the crankshaft until you get the cam just right—at the highest point. Point gap should be .015. Note that some of Tecumseh's two-cycle engines, especially the marine engines, have an "adjustable timed unit" that requires more involved timing. If you have such an engine you will need to follow a different timing procedure—which you can do, but marine engines are outside the scope of this book since they differ in many ways.

If you don't get any spark after installing the new points and condenser, the fault is in the coil assembly. Check the coil visually for cracks, insulation damage, signs of overheating, broken electrical leads where they join the coil; also check the magnets by placing the flywheel upside down on a wooden surface and holding a screwdriver to within an inch of the magnet. The screwdriver blade should be attracted to the magnet.

If you see nothing wrong with the coil, you can check the terminals for continuity, short circuits, and grounds, checking both primary and secondary leads. Of course, you must disconnect these leads when testing. With a continuity tester, test the secondary lead from the spark plug cable to the ground, and the primary lead from the point terminal to the ground.

To summarize: these are the common symptoms of small engine problems.

1. The engine won't start.
2. It starts only with difficulty. It kicks back on starting.
3. It lacks sufficient power. The engine vibrates. It overheats.
4. It consumes too much oil and/or leaks oil.

To summarize ignition problems: Is fuel getting to the spark plug? Determine this by trying to start the engine. If it won't start, remove the plug and look for gas on the plug. Next, use insulated pliers to hold the removed plug near the engine while you crank the engine by turning the starter (or cranking it by hand). If spark flashes across the electrodes of the plug, you are getting ignition. Try a new plug.

If no spark occurs in this test, the fault is not with the plug; it is in the points and condenser, in the magneto, in the air gap of the magneto, in the point plunger, or in a shorted-out cable or switch.

Such faults as kickback, hard starting, vibration, and power loss or noise are most often associated with the equipment driven by the engine. A loose blade or belt—if there is one—can be the cause of several of these faults. Vibration may be caused by any loose bolt or connection, but it can also be caused by internal engine problems, including a bent crankshaft. The engine mounting bolts that hold the engine to the mower deck can work loose and cause vibration. A cracked mounting deck causes vibration, too. Power loss on a mower can be caused by grass buildup under the deck and also by any gears in the equipment being driven by the engine.

CARBURETION

Carburetors are strange devices that mix air and gas at about 15 parts air to one part gas. This mixture is then drawn by the force of engine vacuum into the combustion chamber, below the engine head, where it is exploded by a spark across the electrodes of the spark plug. The fuel mix is compressed by the piston pushing the mixture up against the head, making it more explosive. Carburetors take a lot of beating from vibration, which changes their adjustments, and from dirt, which clogs their arteries. Their moving parts—very few in small engines—inevitably wear, and sooner or later they need overhauling.

The most common problems with small engine carburetors are the simplest to remedy—the screw adjusters and the air cleaners. The adjusters control the air/fuel mix, and the cleaner, usually at the top of the carburetor, often becomes clogged with dirt and won't allow air to flow.

The screw adjusters require only a screwdriver and a couple of turns. With the engine running, adjustments should be made turning the needle valve in until it bottoms, then turning it out two complete turns. These should be done delicately; when the needle valve (screw) seats, don't force it further. That will ruin it. The needle valve is located at the front of most Briggs & Stratton engines, but there is no predicting where you will find it. When you turn it in, the engine should stall or develop a rough idle. As you turn it out, past the two-turn position, turn it very slowly until you get what sounds like the smoothest idle. Though it is best to do this adjustment with the engine fully warmed, if you get the needle valve too far out you may have trouble starting the cold engine the following week. So, don't just turn it. The fuel/air mix is a vital ratio; treat it with respect.

The idle adjustment is less important. If the engine doesn't stall, let it alone. It isn't like a car idle which, if wrong, will cause all kinds of havoc. If it's too fast, that might be annoying; if too slow, it might cause a stall. Generally, it's not an important adjustment. An engine of this size that runs well enough doesn't need an idle adjustment.

Let's look first at common Briggs & Stratton carburetors. The first thing you look at is the air cleaner, which keeps dirt out of the engine, preventing wear, but if the sponge cleaning element is completely dirty it also keeps air out of the engine. That means no go. Briggs & Stratton favors the sponge-with-oil type; others use paper filters as in auto carburetors.

Sponge filters should be taken out, washed, and oiled with engine oil every 25 hours of operation or at least once a summer. If the engine operates year-round it should be cleaned accordingly. The housing that holds the filter atop the engine is held in place by a long screw in the

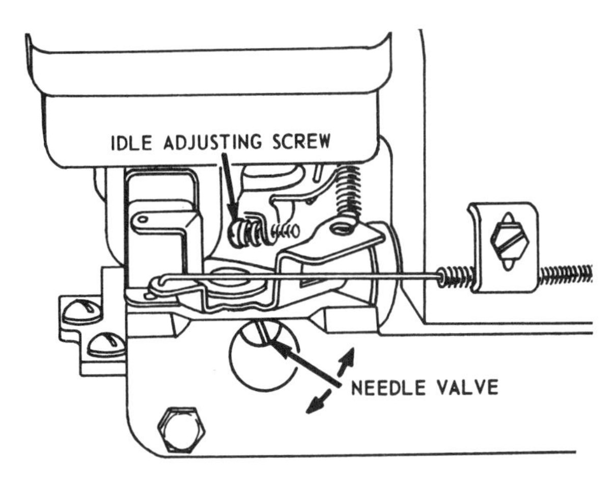

FIGURE 20. Carburetor adjustments on a typical Briggs & Stratton engine.

center. Too much oil in the filter will drip down into the carburetor—a no-no. Thus, you must wring the filter out before replacing it.

In addition to the several shapes of sponge filters one also encounters cartridge-type air cleaners, consisting of a housing, a metallic filter that can be washed and reused. It should *not* be oiled; it should be tapped to remove loose dirt, then washed from the outside in until the water is clear, then dried.

One other type may be encountered; an oil bath cleaner. In this one, the bowl contains engine oil at the bottom. If the oil is dirty, empty it and wash the element.

If the engine idle is unsatisfactory, or starts with difficulty and doesn't run smoothly and with the requisite power, and you have eliminated ignition as the source of the trouble, the carburetor must be examined closely. Note that in many models the gas tank is just below the carburetor and indeed part of it.

One of the most common carburetor ailments isn't the carburetor at all, but the tank; it might, for example, get banged against a tree, disturbing the seal and thus interfering with the right ratio of air to gas. In that case, the best cure is a new tank. Use a straight metal edge across the tank from various angles to test for any tank distortions.

A cautionary note: whenever you remove a carburetor, remember, or write down, the method of attachment of the governor springs and other links, though we illustrate common examples.

Briggs & Stratton uses three basic carburetor styles, which have the names Pulsa-Jet, Vacu-Jet and Flo-Jet. Though they're similar, there are enough differences among them so that they are not interchangeable.

The Pulsa-Jet is the most common Briggs & Stratton carburetor, used on the less expensive engines. To remove the carburetor requires that two gas tank retaining bolts be unscrewed. One is in front, the other on the side. The Briggs & Stratton 82000, 92000, 94000, 110900, and 111900 are the carburetor models. In these engines carburetor and gas tank overlap. It isn't the dismaying conjunction one finds in automobile gas tanks and fuel pumps—a sadistic combination because all fuel pumps fail—and if they are located in the gas tank, rather than in the engine where they belong, it becomes a major task of extrication. The small engine carburetor plus gas tank is an ingenious combination with no service problems. Note that most of these small fuel systems don't use a fuel pump, but pull their gas generated by vacuum pressure from the piston movement in the cylinder in conjunction with a small rubber suction device in the carburetor.

Once you have removed the air cleaner, replacing it or cleaning it, as the case may be, you go next to the carburetor itself.

Removal of the carburetor—gas tank typically requires removal of five thin bolts that come out with a screwdriver, plus two shorter, heavier bolts at front and side that require socket wrench and extension.

An idle control link and a spring from the air vane to the hand control bracket must also be taken off. Notice the way the link and spring are attached. They come off more easily when you remove the screws and bolts from the carburetor and tank and turn the two components so as to ease separation from the links. No great trick.

Separate carburetor and tank, taking care not to damage the two fuel pipes below. Examine the rubber diaphragm for holes, the spring and its seat for wear. The diaphragm should be without holes or other signs of wear. Otherwise it must be replaced.

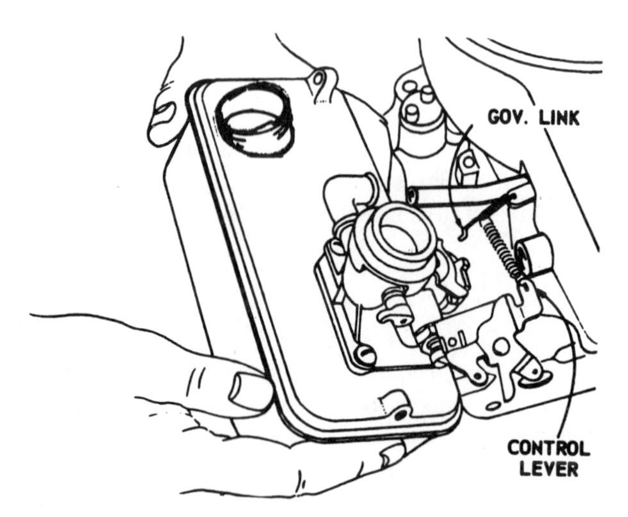

FIGURE 21. Removing the tank assembly from a Briggs & Stratton engine.

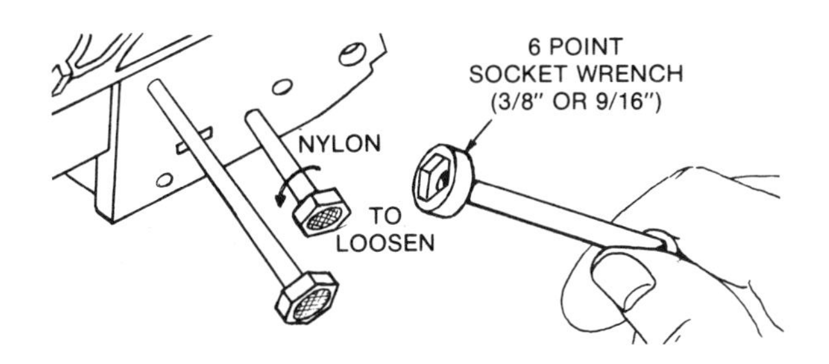

FIGURE 22. Nylon fuel pipes must be kept clean or replaced if they can't be cleaned.

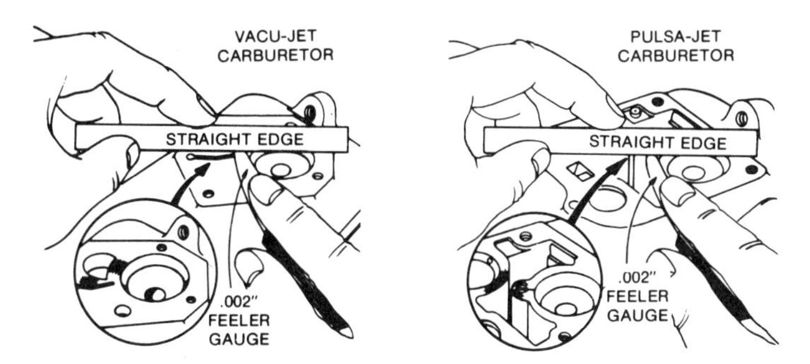

FIGURE 23. Here's how to check tank top flatness.

To determine whether or not the gas tank must be replaced, use a metal ruler or any other flat device and lay it across the tank, pushing it across all possible intersecting surfaces. If the tank is warped, it must be replaced because such a condition will distort the air/fuel mix by adding air randomly, thus rendering correct engine performance impossible.

Note that in checking the tank top for evenness, use a metal or other ruler that is without warp, and a feeler gauge underneath the ruler. A .002-inch feeler gauge should not enter at any point between the ruler and the tank top.

If the tank isn't too far gone—hasn't been mangled from contact with a tree or rock—you can use a Briggs & Stratton repair kit to smooth out the offending part of the tank.

Unscrew the needle valve. It should taper cleanly and without burrs to its pointed end. If it is warped or worn because of incorrect tightening, it, too, must be replaced.

Examine the carburetor for dirt, but notice especially the butterfly choke plate valve. It must open and close easily and precisely. If it sticks, the valve and carburetor must be soaked for an hour or two in carburetor cleaner. Sometimes you can free up a choke valve using spray can carburetor cleaner, dosing it liberally. Try that first. It's easier and less messy.

The key to small engine carburetor performance is the operation of the automatic choke, an ingenious device in these small engines. First, the Briggs & Stratton.

The choke operates basically from engine vaccum rather than from heat, as with car chokes (which also use electronic sensors and other gear). The choke valve is operated by a link, and a spring that goes down to the diaphragm. The spring holds the valve in closed position when the engine is not running—that is, before you start it. The intake stroke, at starting, creates vacuum that pulls open the choke by pulling down on the diaphragm. If the spring is too weak to keep the choke closed for starting, you need a new spring. When engine speed decreases, the choke is operated by the choke system to enrich the mix—it is partially closed.

Testing the choke system can be done visually. Remove the air cleaner to expose the choke butterfly valve. It should be completely closed when the engine isn't running. Move the accelerator hand control to the "stop" position; the throttle lever, which abuts the hand control bracket and is attached to a spring (and has a setscrew adjustment), should be in a closed position. When you use the starter, the choke valve should flutter open and shut.

Then, if the engine starts, let it run for a couple of minutes at operating speed, with the fuel tank at least half-full. Open the needle valve several turns until the engine coughs, then close it until the engine threatens to stop, then adjust it midway. Allow the engine to run at idle for three to five minutes. Now, close the needle valve, and the engine should stop from too lean a mixture. If the engine continues to run, that means a fuel leak is occurring—probably through the diaphragm, but it

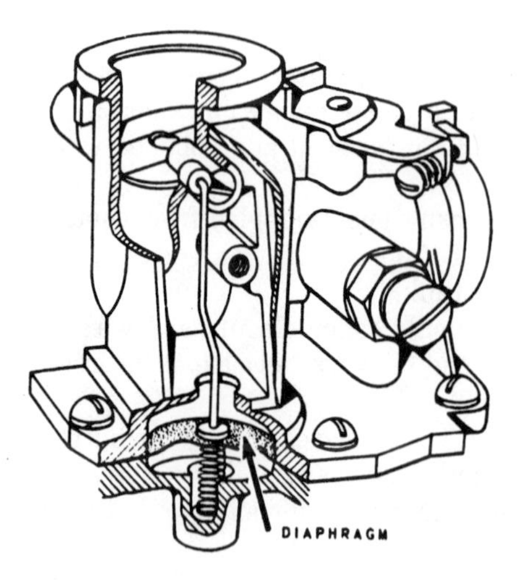

FIGURE 24. A cutaway view of the choke system of a Briggs & Stratton engine.

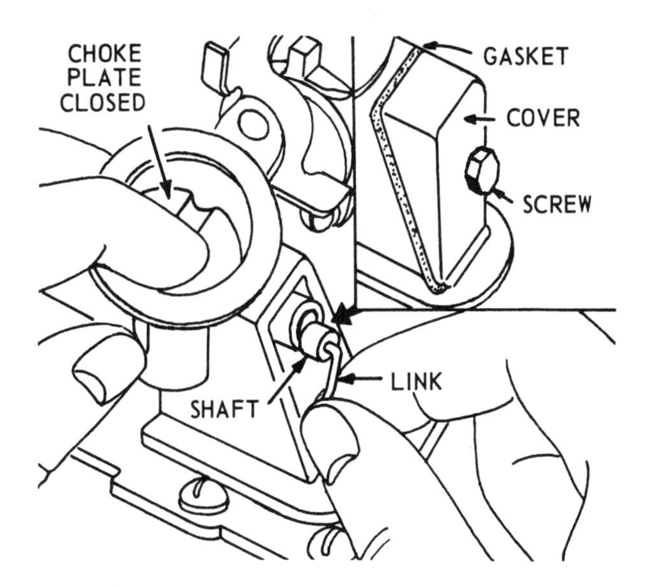

FIGURE 25. Choke link, gasket, cover, and screw.

may also occur because of a warped tank top or a loose needle valve fitting.

If the choke valve reacts correctly during starting and running, and you continue to have operating problems with the engine, the fault is elsewhere in the engine.

When you take the carburetor off the tank, exposing the diaphragm, also remove the choke link cover as required. If the diaphragm is undamaaged it can be reused. Measure the spring in the choke system; the Pulsa-Jet spring minimum length is $1\frac{1}{8}$ inches, maximum is $1\frac{7}{32}$ inches. In the Vacu-Jet, the minimum is $\frac{15}{16}$ inch, maximum 1 inch. On Model 110900 and 111900 the minimum is $1\frac{5}{16}$ inches, maximum $1\frac{3}{8}$ inches. If lengths differ you have to replace both diaphragm and spring.

It's tricky to reassemble the carburetor, tank, diaphragm, and automatic choke. Treat the choke spring with care because it is calibrated; push it into its base in the diaphragm. Then turn the carburetor upside down, put the diaphragm on top (that is the bottom), where it belongs, while you get the choke link through its hole. The Pulsa-Jet requires the pump spring and cap to be in the special pump well; the other types do not. Place the tank down *gently* on the carburetor, guiding the choke spring into its position. Hold the carburetor and tank together, turn them right side up, and finger-tighten the mounting screws into the tank. Close the choke valve and connect the choke link with the choke shaft. Move the choke plate to an over-center (open partway) position. Tighten the carburetor mounting screws in an opposite sequence. The choke should now be closed. If it is not, the choke spring may not be correctly assembled. Put the choke link cover and

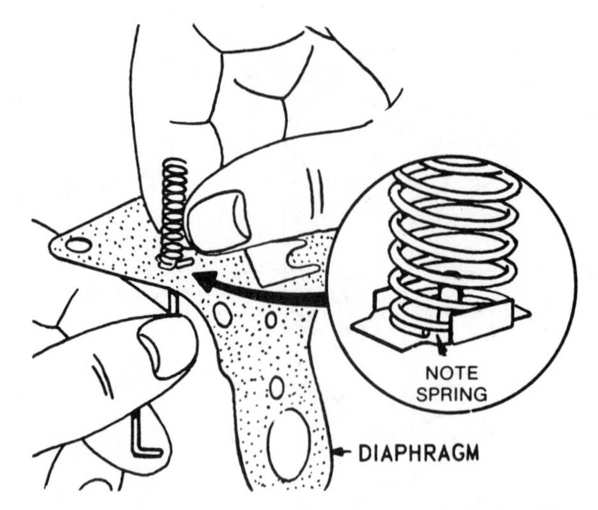

FIGURE 26. Here's the way the choke spring fits into the seat and the diaphragm.

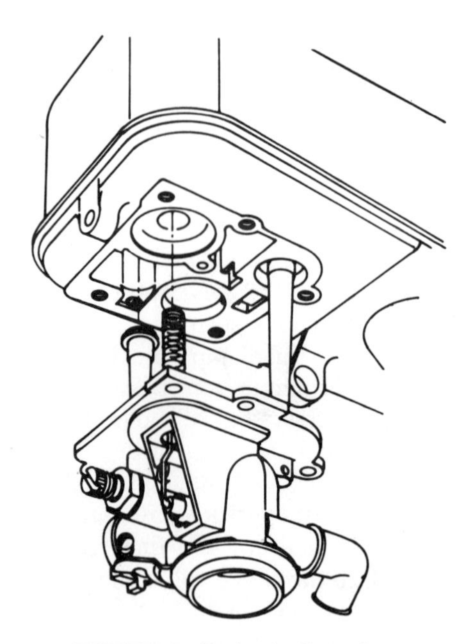

FIGURE 27. Putting it all together.

gasket back. Start the engine and warm it up. If it won't start, turn the needle valve in all the way (gently), then back it out 1½ turns. That should start it. If not, prime it with a bit of gasoline poured into the throat of the carburetor.

To adjust the carburetor, turn the idle adjusting screw until you get an idle speed that allows correct acceleration. Turn the idle adjusting screw to a point where the engine can be both turned off with the speed control lever and accelerated without stalling. You will also have to get the needle valve at the best adjustment in conjunction with the idle speed, though the 1½-turn adjustment is bound to be close to the mark.

You encounter Vacu-Jet carburetors on Briggs & Stratton engines 82500, 92500, and 94500. Pulsa-Jet is on 82900, 92900, 94900, 110900, and 111900, as well as earlier models. There are some fastening variations from year to year, but a little study will quickly zero you in on the correct sequence.

The Minlon carburetor, (plastic base) on model series 92500 and 94500 (Briggs & Stratton), has a single fuel pipe with a press fit design—it pulls in and out rather than being threaded. It snaps in and out with some force; you pull it out with a pair of pliers and install it that way. Minlons cannot be soaked for more than 15 minutes; otherwise, they dissolve.

In general, these small engine carburetors are models of great ingenuity and simplicity. In some ways they are harder to deal with than automobile carburetors in that their very simplicity can be deceptive—for example, if you don't get the ingenious choke spring and diaphragm setting exactly right you court trouble. The major things that go wrong with auto carburetors don't happen with these simply because they don't have the quantity of components that can go wrong.

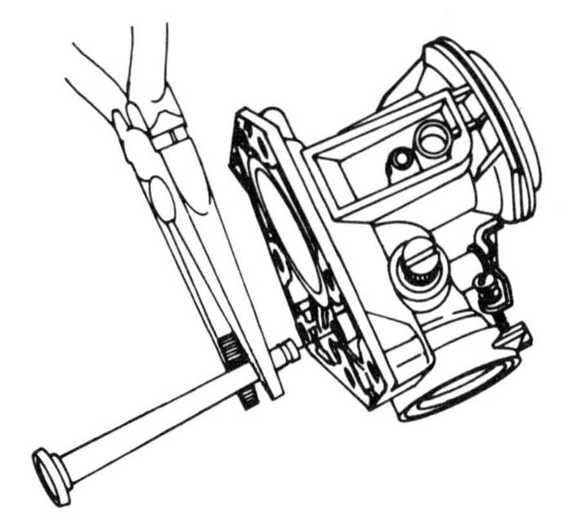

FIGURE 28. Fuel pipe replacement on the Minlon carburetor.

Briggs & Stratton carburetors have a wide variety of small differences in fastening, in disassembly, and in the most simple removal procedures. Some have been noted, others follow.

Model series 110900 and 111900 hide a carburetor mounting screw under the choke valve. It takes a Phillips screwdriver and you have to open the choke butterfly valve all the way to get the screwdriver on the mounting bolt.

On the 82000 and 92000 model series, some carburetors have a small poppet valve (pressure-opened) located on the choke plate, which is part of the "Choke-A-Matic" automatic choking system. To remove the choke parts you must first disconnect the exterior choke lever spring, then pull the nylon choke shaft sideways, thus separating the choke shaft from the choke valve. Some of these models' choke valves are heat-sealed to the shaft. You can loosen the shaft by pushing a knife along the edge of the shaft to break the seal. It isn't necessary to reseal this joint. Install the choke valve so that the poppet valve spring is visible when the choke plate is fully closed.

On Vacu-Jet carburetors the fuel pickup tube has a check ball in it that must operate freely. Clean the screen, and if the cleaning operation doesn't allow the check ball to move freely, it is necessary to replace the pipe.

On the Briggs & Stratton carburetors known as Zinc (from the model series 82000, 92000, 92500, 94500, 110900—the Minlon—and 111900) using the Pulsa-Jet or Vacu-Jet types, you may encounter needle valve assemblies with a lot of little washers and rings. Use new O rings, gaskets, washers, and diaphragms. If you are cleaning the carburetor body it will be necessary to remove these needle valve assemblies. Make a note of their order, and note also the amount of tightness involved on the needle valve seat and its relation to the little flat in the carburetor body. It isn't very tight. The flat on the valve seat lines up

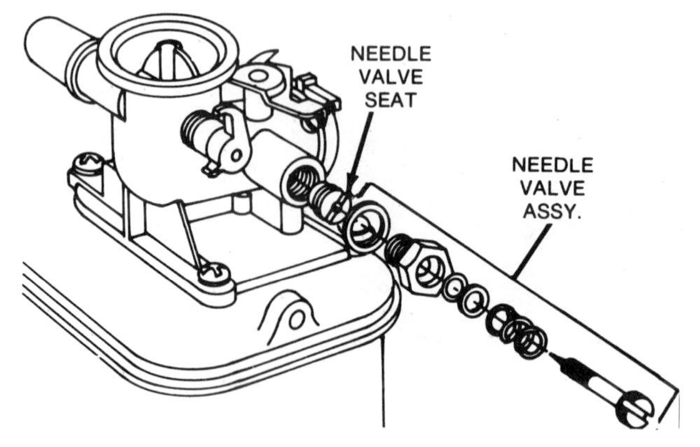

FIGURE 29. Needle valve assembly on Zinc body and Minlon body carburetors are slightly different.

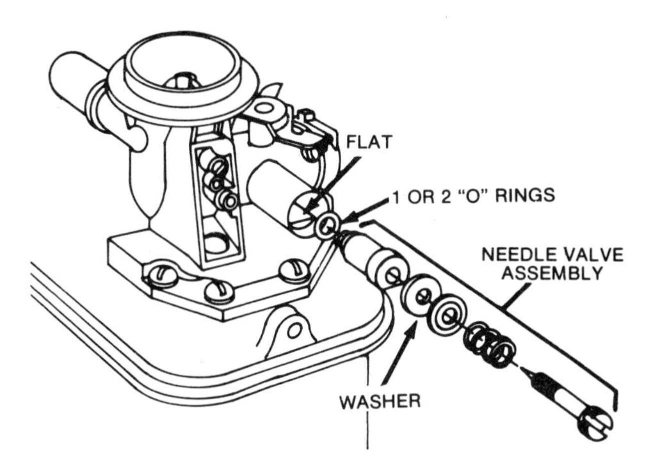

FIGURE 30. Minlon needle valve assembly.

with the flat in the carburetor. Note that an O ring goes in a groove in the throttle bore. Briggs & Stratton suggests dousing an oil film on every new O ring.

Vacu-Jet carburetors have a gasket between tank and carburetor; Pulsa-Jets have the diaphragm that serves as a gasket. If you crease or otherwise distort either gasket or diaphragm in the process of assembling the two components (tank and carburetor), you will almost certainly get a balky engine—or worse, one that won't run at all.

When it comes time to put the carburetor and tank assembly back together on the 82000 and 92000, the fitting of all the controls can be tricky. Align the carburetor with the intake tube and breather tube grommet. Hold the choke lever so that it doesn't catch on the control plate. Watch the O ring in the carburetor when fitting into the intake tube—don't squeeze the O ring.

When adjusting most of these carburetors you need half a tank of gas. You should also do the adjusting with the carburetor air cleaner and its retaining bolt in place. If you don't, you will have an engine that works properly only without the air cleaner on it, in which case the carburetor will quickly fill with grass clippings, stray dirt, and bugs.

These caveats may make it seem that small engine adjustment is worse than adjusting your car engine. To some extent that is true, but these engines are remarkably forgiving—they are "user friendly," as computer jargon puts it. You can't get away with murder, but bad adjustments won't be fatal, and if you make a bad mistake the correction is usually easy.

On the 82000 the breather tube and fuel intake tube thread into the cylinder. The fuel intake tube bolts to the engine on 92000, 94000, 110900, and 111900. Examine the gaskets at the fittings; damaged gaskets allow air leaks and dirt entry. These are no-no's.

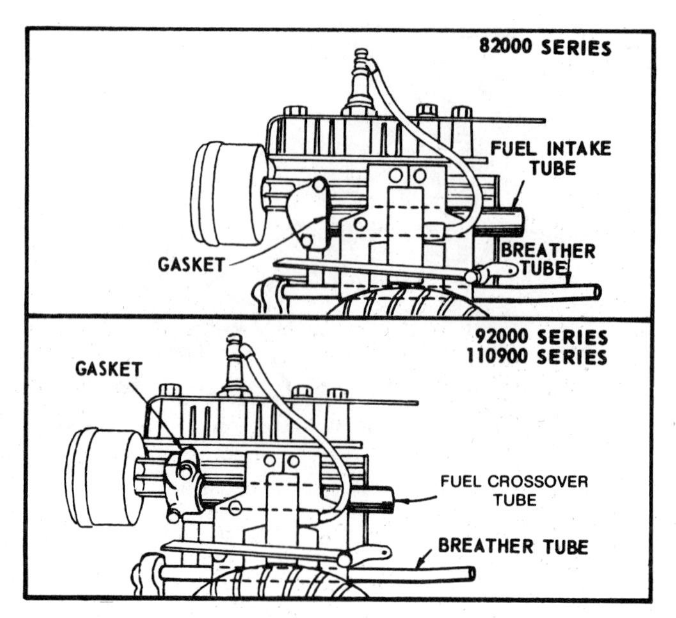

FIGURE 31. Details of breather tube and fuel intake tube on differing Briggs & Stratton carburetors.

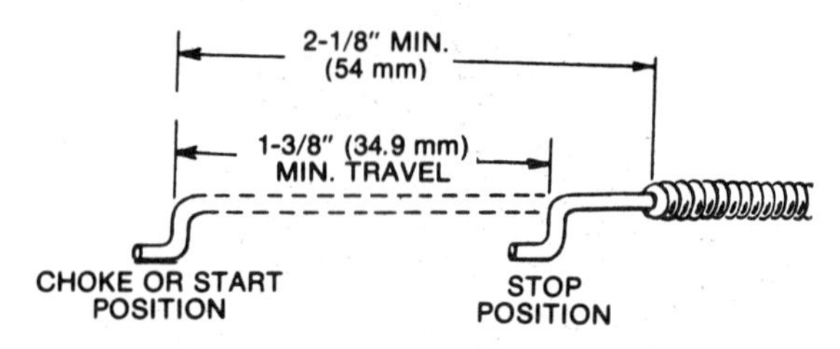

FIGURE 32. Choke-A-Matic adjustment.

The Choke-A-Matic adjustment is a standard feature on Briggs models 82000, 92500, 92900, and various submodels of these generic numbers. The travel distance of the remote control cable out of its casing is 1⅜ inches when you move the control lever from the stop to the choke or start position.

To install a new remote control assembly on the Choke-A-Matic: Remove air cleaner and move the control lever midway between idle and fast. Mount the remote control in the prescribed position on the handlebars. It should be set to the "fast" position. As shown, lever "A" on the carburetor should just touch the choke shaft at "B." Move the wire casing "D" until the correct position is obtained, then tighten the set screw "C," after which you check the control settings with the engine running. The reason you need to put on a new control assembly is that the parts simply wear out—the wire and its casing eventually stick, and the control lever breaks or wears out. Note that you can make these control devices last far beyond normal life by oiling and greasing them periodically. You can put engine oil on the casing each season. It will soak through and lubricate the wire, extending its life. The control lever can also be lubricated to good effect. Don't hesitate to use an oilcan on most moving parts of these small engines and controls. It's also a good idea to put grease on the control plate at the carburetor end, where the casing and wire attach.

Some Pulsa-Jet carburetors have a stop switch ground wire attached to the carburetor. Remove the wire before you unbolt the carburetor.

Other Pulsa-Jets may have a special spiral in the carburetor bore. The spiral is a press fit, and if there is any need to remove it you can pry it out with a pliers and screwdriver. Put the carburetor in a vise, hold the pliers and spiral with one hand, pry up the pliers with the screwdriver against the vise. The spiral is part of the design to mix air and fuel in the carburetor bore.

Pulsa-Jets come in several different shapes and devices. The Choke-A-Matic linkage that goes with these carburetors, and connects the choke and choke controls, has slightly different connecting and fastening details from model to model. A little study will enable you to puzzle out these details, but here are some hints.

When you work on a Pulsa-Jet whose Choke-A-Matic controls are mounted on the carburetor assembly—which could include snowblowers, lawn mowers, or other devices—the procedures differ from the remote control (or handlebar) mounting types: move the speed adjustment lever to the choke position and check the choke slide at the front of the carburetor. It should close, fully. If it doesn't, bend the choke link until it does, to take up the slack. Bend the link about halfway down, bending it upward. Use a pliers. The same speed adjustment lever must make a good contact against the stop switch when it's in the stop position. The test is purely pragmatic—if it stops the engine, it's doing its duty.

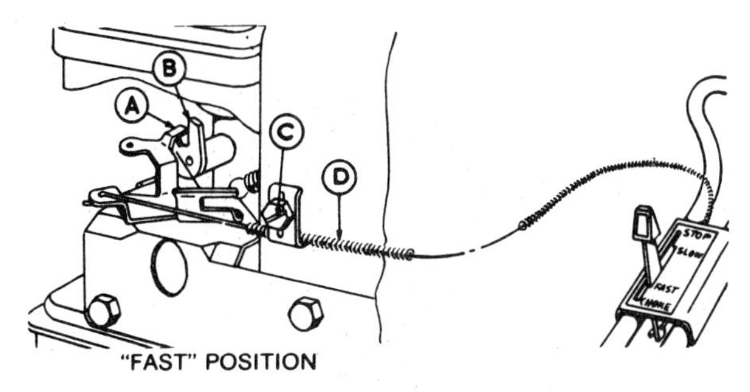

FIGURE 33. Typical Choke-A-Matic remote control.

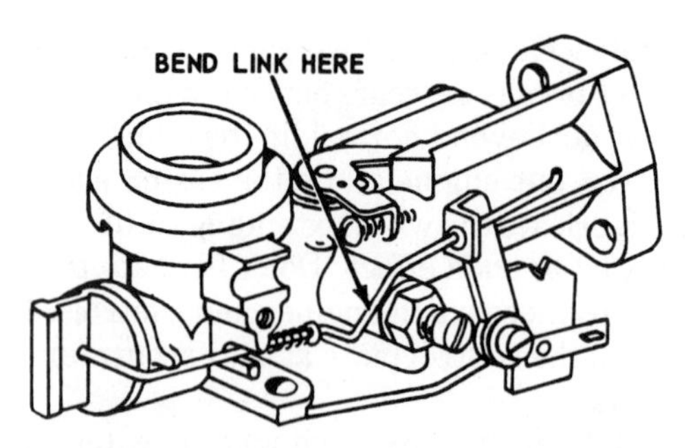

FIGURE 34. Adjustment on a new control assembly (Pulsa-Jet).

In models 100900, 130900, and 131900, with either manual or remote control (manual being at the carburetor, remote on handlebars or elsewhere), check the lever on the control plate at the carburetor. If it's remote control, the lever has a loose fit; but if it's manual control it is by press fit. To check the lever action, move it to the left until it fits snugly into run detents. Lever "B" should just touch the choke lever at point "C." If it does not, loosen the screws "A" and move the control plate to right or left until the lever does just touch the choke lever at the prescribed point "C." Then tighten screws.

Vacu-Jet carburetors by Briggs & Stratton have their own peculiarities. These carburetors have a check ball and fine-mesh screen in the fuel pipe. The screen must be without dirt or debris and the check ball must move freely. Fuel pipes, as noted before, are made of plastic or brass. If plastic, they can be unscrewed with a socket wrench. Brass pipes must be pried out, using either a vise pliers or vise. To replace the brass pipe, tap it in until it projects 2⁹⁄₃₂ inches to 2⁵⁄₁₆ inches from the carburetor gasket surface. You can also press the pipe in, using a vise, but if you tap it be careful about bending it or distorting the pipe ends.

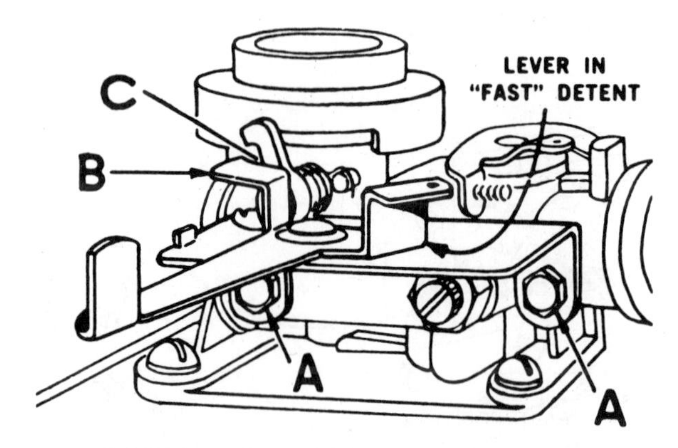

FIGURE 35. Choke-A-Matic linkage adjustment.

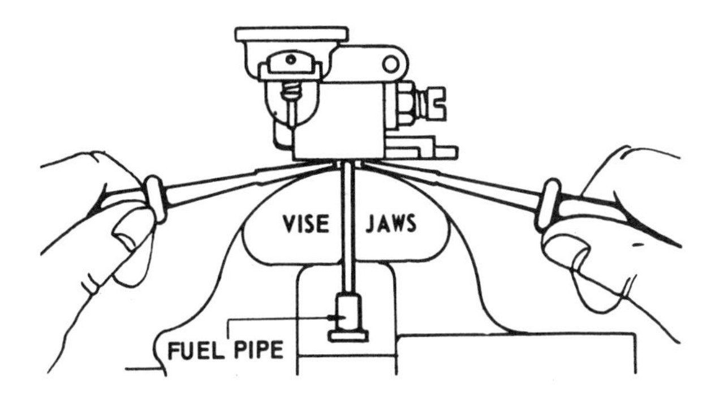

FIGURE 36. Vacu-Jet brass fuel pipe removal.

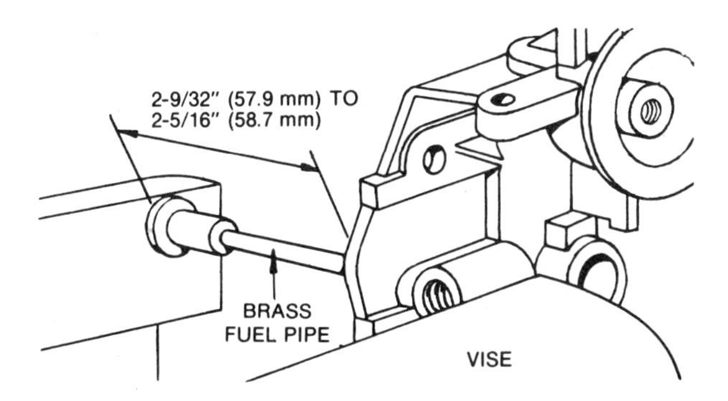

FIGURE 37. Replacing a Vacu-Jet brass fuel pipe.

The one-piece Flo-Jet carburetor comes in a small and large version. The small version differs little from the larger one. Both are automobile-type carburetors, with floats to control the amount of gasoline flowing through, rather than the gravity or compression-demand schemes of the Pulsa-Jets. The float moves up and down as gasoline enters and leaves the float chamber in response to engine demand and the pull of compression. A needle valve controls gasoline flow.

The Flo-Jets are automobile-type carburetors, but vastly simplified; nobody will confuse the two. For purposes of adjustment, Flo-Jets have more nuances of agreement than do the Pulsa-Jets. The Flo-Jet has high-speed and idle-speed adjusting needle valves; the smaller version has these adjusters opposite each other on top of the carburetor, whereas the larger carburetor adjustments are on top and bottom.

In the unlikely event that a carburetor overhaul of one of these Flo-Jets is necessary, the problems are similar to an older auto carburetor overhaul. You have to remove the float and float valve assembly, check the valve and its seat for wear, examine the float for leaks, and adjust the float's up and down movements precisely so as to ensure correct fuel flow.

The float controls fuel flow; if it isn't set exactly, the engine won't know how to behave. It won't start, or, if it does, it won't run properly.

Remove the needles and inspect them for damage. Any bend, groove, burr, or other wear means you must replace the needle.

You can buy a carburetor repair kit that will have new gaskets, needle valve assemblies, and other parts. You don't have to use every new part, but every old one you replace puts the carburetor that much ahead of the game. In general, you will need to replace the float needle valve, if it has had lengthy use, and all the gaskets, possibly the idle and high-speed mixture needles, and any O rings.

If you replace the float needle valve (after checking the float itself for leaks by shaking it near your ear), you have to set the float level. The question is, when do you replace the needle valve itself? Examine it carefully in a strong light to look for wear on the Teflon tip. A metal tip is examined for the same defects. Don't replace it unless you can see such wear; you may not improve matters.

If you do replace the needle valve, it will be necessary to set the float. You measure the setting with the body gasket in place and with the float valve and float installed. The float must be parallel to the body mounting surface. Don't press on the float to get it where it belongs; that will ruin the needle valve. There is a tang at the attaching end of the float that you bend, using a needle-nose pliers, to get the float in the correct relationship to the mounting surface. When it is closed, the float should be parallel to the surface.

Flo-Jet carburetors also come in small, medium, and large two-piece configurations. They closely resemble the ones we've just been talking about. They add a few procedures, but not much that is different.

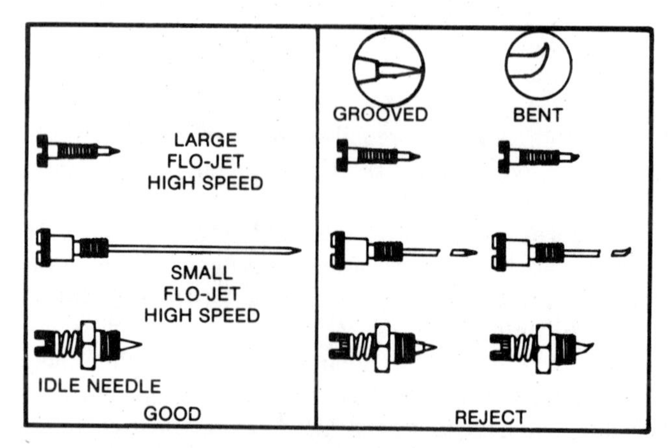

FIGURE 38. Good and bad needles.

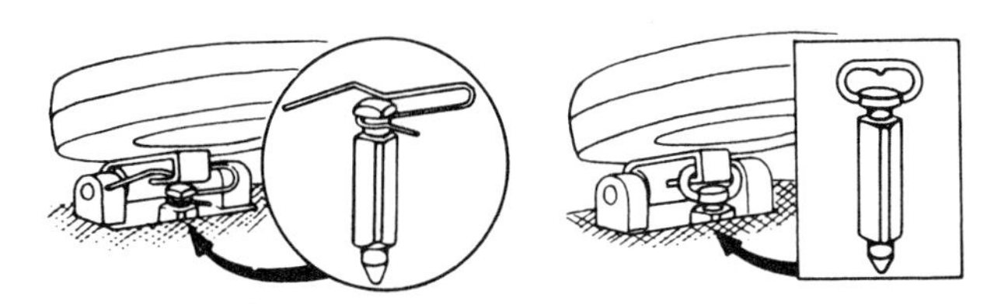

FIGURE 39. Needle valve fastening variations.

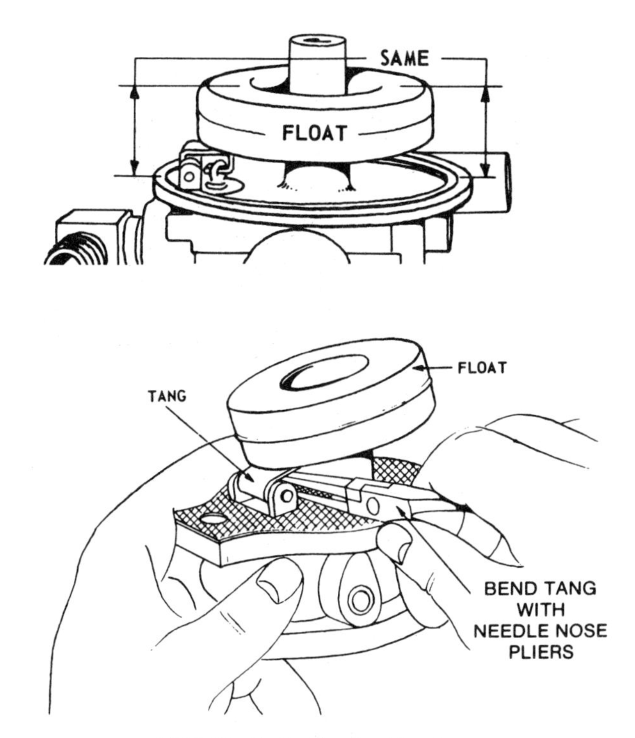

FIGURE 40. Setting float levels.

The float and needle valve replacement and adjustments are the same.

The Flo-Jet carburetors, with their autolike float and needle valve assemblies, are subject to throttle shaft and bushing wear. The float valve seat is also a press fit, which must be replaced if it develops a leak. These are machining tasks that require exertions different from screwdriver, pliers, and socket wrench manipulations. Be warned, if you take them on, that what you do is different and you might not want to tackle these two jobs. However, a few simple routines are required. Here they are.

First, the throttle shaft bushing must be replaced when it gets loose. Looseness between the shaft and bushing should not exceed .010 inch. Check the wear by putting a short bar of some sort on the upper carburetor body and then measuring the distance between the bar and shaft with a feeler gauge. Hold the shaft down, then hold it up; if the difference is over .010 inch you must either replace the bushing or do more fundamental work.

To replace the bushing, put a ¼-inch × 20 tap (or an E-Z Out) in a vise. Turn the carburetor body so that you can thread the tap into the bushings sufficiently to pull them out. Put new ones in with a vise—don't pound them. Put the throttle in to make sure that it will turn. When removing the throttle shaft, note that it is held in place with a pin. With a thin punch, drive the pin out, then pull out the shaft.

For complete disassembly of these small Flo-Jet carburetors—if it should ever become necessary—you begin by taking out the idle valve, then loosen the high-speed valve packing nut and remove the packing nut and needle valve together. On the medium and large sizes, remove the high-speed valve assembly as a unit. The nozzle, which threads diagonally through the carburetor, must be unscrewed carefully and with either a special screwdriver or a small one. You want to avoid damaging the threads. The nozzle must be removed; otherwise, you can't separate the top from the bottom of the carburetor, without ruining the nozzle. Of course, you can buy a new one if you ruin it. You can also ruin the carburetor body in the process, which is costlier than the nozzle.

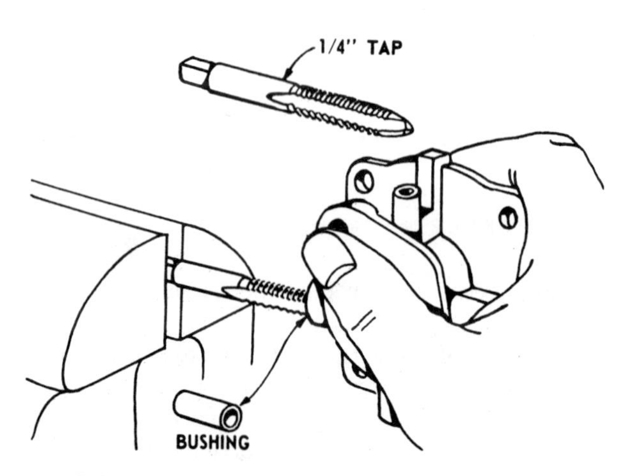

FIGURE 41. Replacing throttle shaft bushings.

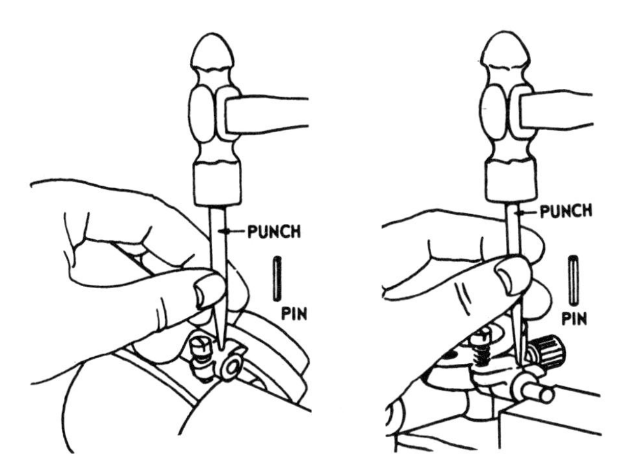

FIGURE 42. Driving out the throttle shaft bushing pin.

Take out the float and needle valve; then, using a wide-bottomed screwdriver, unscrew the float inlet seat. Only a wide-bottomed tool will fit across the opening of the seat.

Some of the float seats are pressed in; they must be pressed out, in a vise. More disassembly is possible but almost never needed, and involves the choke plate and shaft.

Press fit float seats must be removed with a self-threading screw that you can buy from any Briggs & Stratton store—the part number is 93029. Clamp the head of the screw in a vise, and thread the carburetor body as described earlier. You turn the body over the screw threads, thus drawing out the float valve seat—as with the bushings before. Place the new valve seat into the carburetor body, making sure that you get the right seat. There are two—one for the carburetor, stamped with the letter "P" on the flange, indicating a fuel pump present in the system; and the other without such a letter. Using the screw and old seat as a driver, press the new seat flush with the body. Don't press it below the body surface.

One other type of Flo-Jet may be mentioned, the crossover. This one, found on the series 253400 engine, has an integral fuel pump. The rest of it is somewhat similar to other Flo-Jet carburetors. The pump is one of those ingenious damping diaphragm devices that appears in embryo form in the simpler tank-type carburetors of the Pulsa-Jets. It isn't a pump that is driven by mechanical or electrical force, as with auto fuel pumps, but by engine vacuum pressure. This Flo-Jet type is found on larger engines.

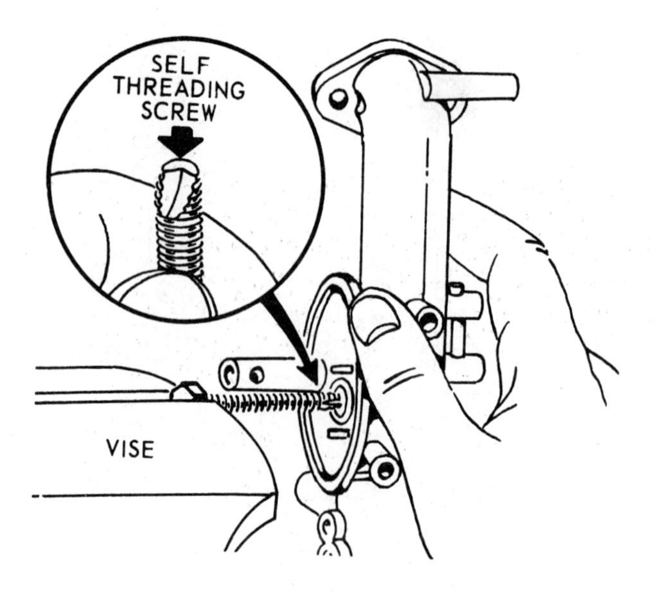

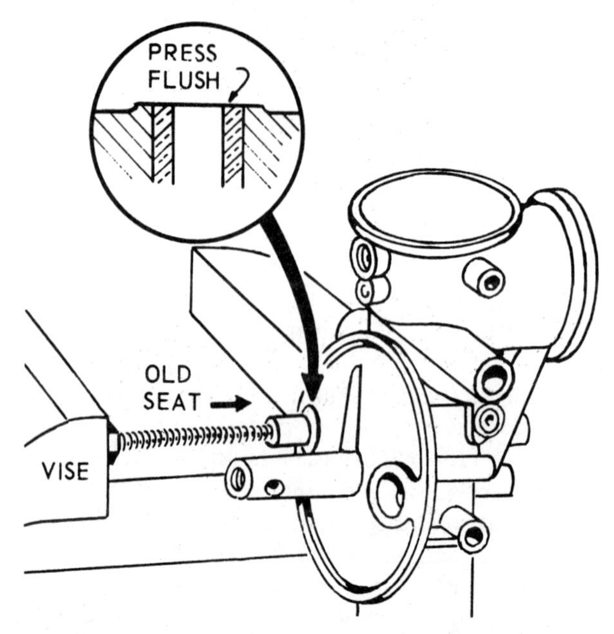

FIGURE 43. The float valve seat is removed and replaced as in the top drawing—you have to pull the old one out and press the new one in.

If the fuel pump develops a gasket leak or other defect, and doesn't produce gas for the carburetor, it will have to be taken apart and repaired. Mostly the gasket and springs and diaphragm get replaced, with some cleaning up of the surfaces. A repair kit for the fuel pump, consisting of all the wearing parts—springs, gaskets, diaphragm, and so on—is available and should be used if the pump leaks, distorts fuel/air mix, or fails in other ways. Remove the pump from the carburetor by unscrewing the four screws from the cover. Remove the fuel supply line from tank to pump and plug the fuel line, or, if a valve is present, turn it closed. Clean the pump before installing the new parts.

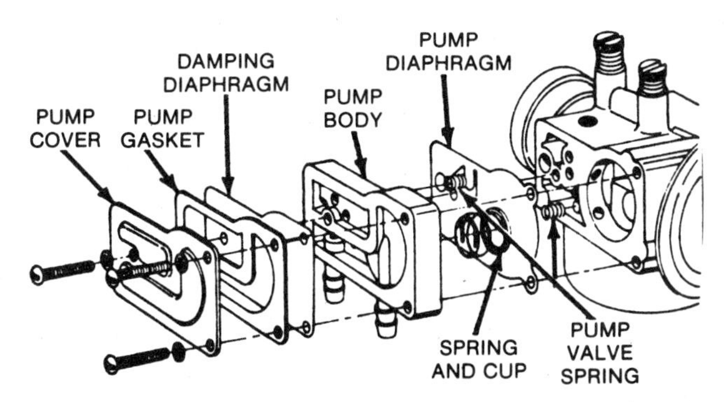

FIGURE 44. A Flo-Jet crossover model by Briggs & Stratton has a special fuel pump, shown here disassembled.

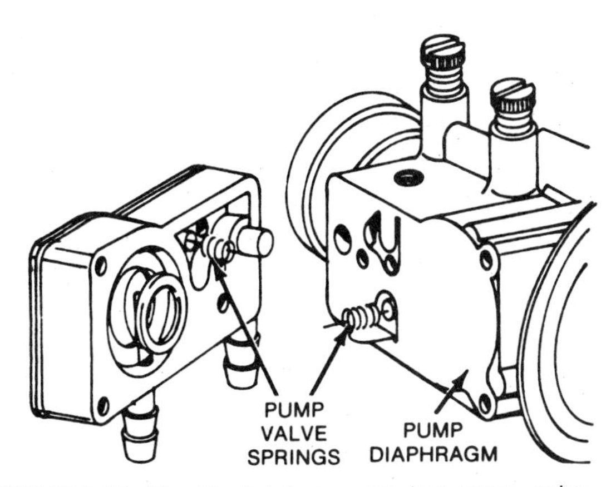

FIGURE 45. The Flo-Jet fuel pump has pump valve springs like simpler models, shown here.

TECUMSEH CARBURETORS

Tecumseh engines are found on such popular lawn mowers as Toro and in many other applications. Their carburetors are the same basic types as those previously encountered, using diaphragm vacuum pressure types as well as the float, autolike type. Service and repair procedures are similar, even though carburetor details are different.

Most carburetor problems on lawn mowers can be cured with adjustment. These small engines vibrate so that the needle valves are eventually changed and their adjustments are skewed. If you can get the engine to run, let it warm up thoroughly. Then put the throttle or speed control on the "run" or "fast" position, and adjust the idle screw clockwise or counterclockwise until the engine runs as smoothly as possible. With the speed control in the idle or slow position, adjust the idle speed until the engine runs its smoothest. See the facing page for a list of some potential carburetor troubles.

Tecumseh lists trouble spots and things to do about them. The atmospheric vent, in the mounting flange immediately under the plate, must be free of dirt. The throttle return spring can get out of whack from dirt or paint, causing the engine to falter. The fuel bowl drain is the engine's way of clearing the carburetor of dirt and gasoline when you store the engine. It should be examined for leaks. An internal rubber seat may require replacement. The choke lever is positioned by its stop spring, which may lose tension and require replacement. If you see gas leaks around the carburetor, the fuel inlet fitting could be loose or cracked. If it was tightened excessively it could have cracked the carburetor body. We will look at these and other possible trouble spots.

Tecumseh small engine carburetors are identified by a model number and code date, stamped onto the carburetor body. This makes the ordering of parts simple and unerring. Both the types that use floats and the diaphragm types do the same thing—mix air and gas—but the diaphragm carburetor can be operated in any position, and so it has applications where the device must be used in tricky angles—for example in a power saw where it is sometimes necessary to cut upside down. A float type wouldn't work here; turned upside down, the float would open completely and flood, and the gas would slosh out. The diaphragm carburetor works on manifold pressure on the engine side, and atmospheric pressure on the carburetor side. As engine pressure decreases in response to engine cycle demand, the diaphragm moves against the inlet needle, lifting the needle from its seat. This allows fuel to flow through the inlet valve and maintain fuel level. Without correct fuel level you can't get the right air/gas mix, and without that you can't get engine firing.

TROUBLE
Carburetor out of adjustment
Engine will not start
Engine will not accelerate
Engine hunts (at idle or high speed)
Engine will not idle
Engine lacks power at high speed
Carburetor floods
Carburetor leaks
Engine overspeeds
Idle speed is excessive
Choke does not open fully
Engine starves for fuel at high speed (leans out)
Carburetor runs rich with main adjustment needle shut off
Performance unsatisfactory after being serviced

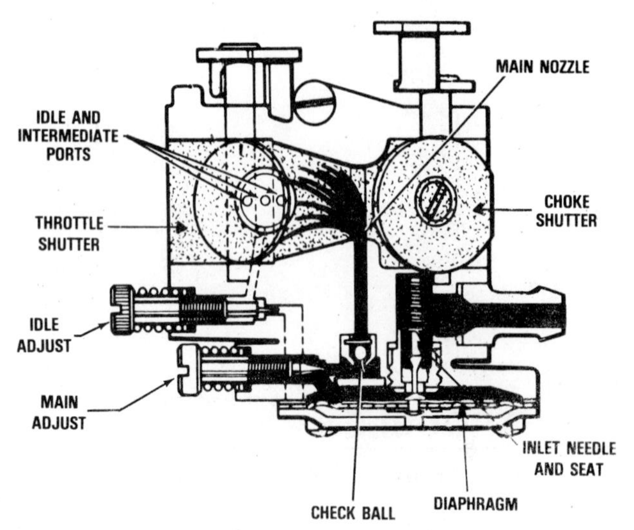

FIGURE 46. A schematic drawing of a Tecumseh pressure-diaphragm carburetor.

When you take apart any Tecumseh carburetor you may find nylon check balls, especially in some of the diaphragm types. These must be replaced, as they cannot be cleaned. Check balls regulate flow without measuring it—they smooth it out.

Welch plugs, used mostly in the choke valve, but also in the fuel system, are recommended by Tecumseh to be removed and replaced whenever the carburetor needs rebuilding. One hesitates to make a hard-and-fast rule about this, but the passageways controlled by these plugs can't be cleaned completely unless you can get at them. To remove these plugs requires that you use a sharp object to pierce the plug and pry it out. Warning: Don't let the pointed object carve up any other part of the carburetor around the plug. The kit that you buy when rebuilding carburetors will contain new plugs, including both the welch and other kinds of plugs found in these carburetors. To replace the welch plug, after you clean out the surrounding territory it covers, install the plug into its receptacle with the raised portion up. Use a punch or a socket equal to or greater than the size of the plug. Flatten the plug, but don't dent or drive the center of it below the top surface of the carburetor.

Examine the throttle lever and choke plate before you take anything apart. You are looking for wear and damage. These parts should fit trimly and tautly, but there are no special specifications indicating wear. If the various moving parts, including the choke lever and shaft and its linking places and bearing points, move freely, you are home free. If they bind, because of wear, you can sometimes correct them

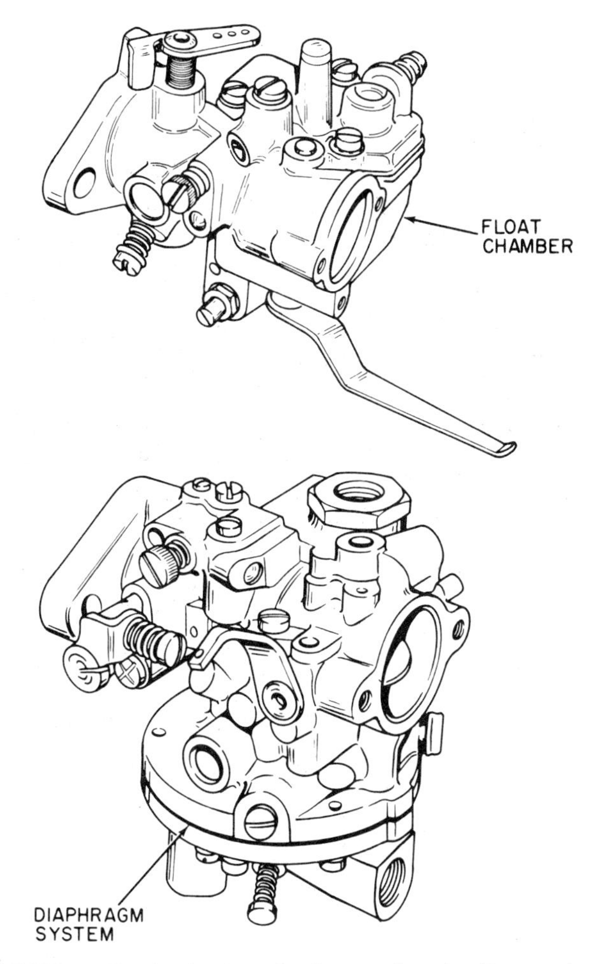

FIGURE 47. Two basic types of carburetors found on Tecumseh (and Lauson) carburetors are the float chamber and the diaphragm system.

with cleaning and greasing. Binding caused by bent links or parts can be cured only with new parts.

The throttle lever and throttle plate—if worn, bent, or otherwise damaged—can be replaced, though it is unlikely to need such surgery. To do it requires removal of the screw in the center of the throttle plate. This allows you to pull out the shaft lever assembly. It is more likely that you need to do it for cleaning purposes. When putting it back together, the lines on the throttle plate must face out in the closed position. Position the plates with two lines at 12 o'clock and at 3 o'clock. If your throttle plate has only one line, the line should be in the 12 o'clock position. If there is any binding, loosen the screws and adjust the throttle plate's position.

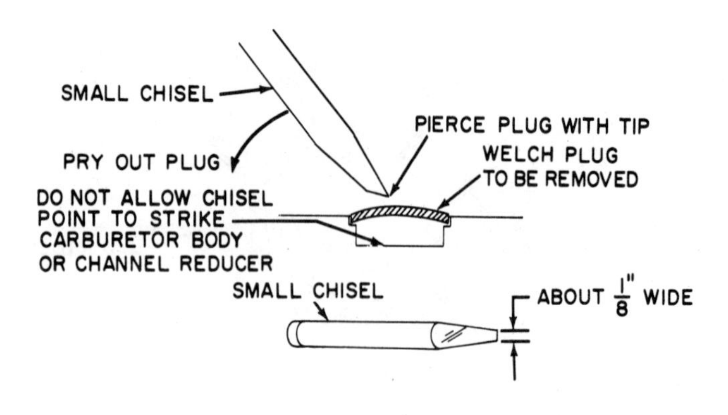

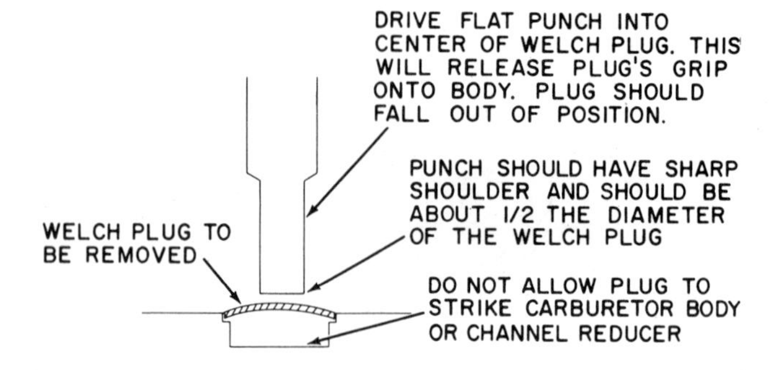

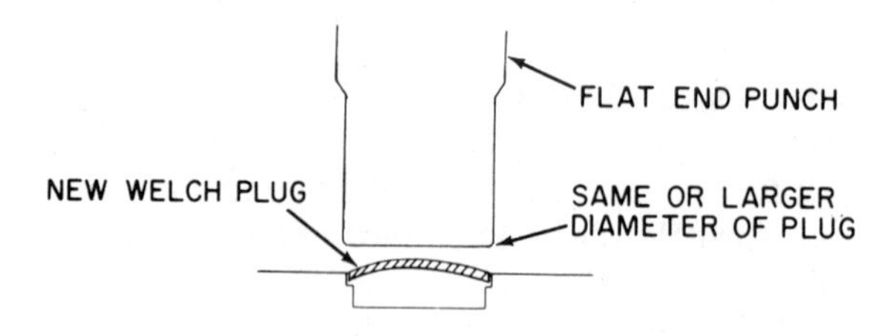

FIGURE 48. Welch plug service on Tecumseh (and other) carburetors consists of removal as shown and pressing the new one in.

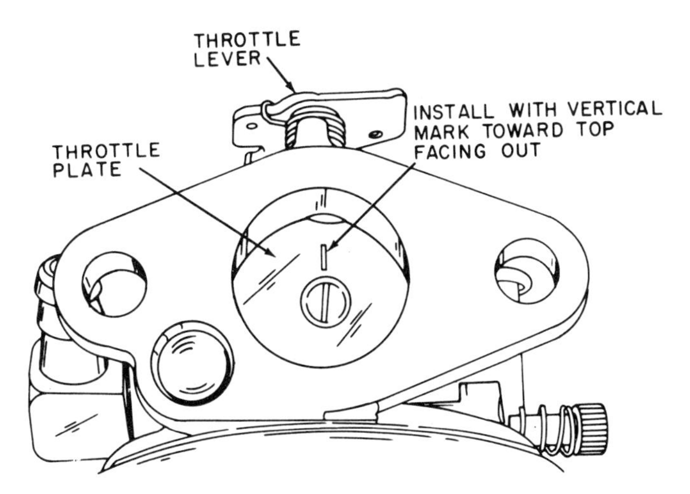

FIGURE 49. Throttle lever and throttle place or shutter need work when they bind or become loose.

The choke plate (or shutter, as it may also be called) is positioned in the air horn of the carburetor (that's the main air cavern that you see when you take the air cleaner off) so that the flat surface of the choke is down. Because choke plates can operate in either direction, it is important to note the correct direction of the choke plate movement.

Take out the idle adjusting screw and check it for damage on the tapered point. Such damage is cause for replacement.

When you install the idle adjustment screw on the 2.7-h.p. Tecumseh engine, the carburetor must be in an upright position. Otherwise the metering rod could be damaged, to say nothing of the carburetor casting.

The high-speed adjusting screw should also be examined for damage. This adjustment screw also has a washer, and an O ring that fits into the retainer nut. Not all Tecumseh carburetors have high-speed adjusters.

The fuel bowl retaining nut should be examined if the engine doesn't idle properly. It contains a passage through which fuel goes to both the high-speed and idle fuel systems. Any dirt or buildup needs correcting—cleaning out.

Examine the fuel bowl for dirt and corrosion. The fuel bowl flat surface has to be on the same side of the carburetor as the fuel inlet fitting—opposite the float hinge pin—to allow the float its necessary movement. To remove the float, pull out the hinge pin (use a needle-nose pliers or your fingers), and pull the float off. This also lifts the needle out because it is held by a spring clip that hooks on the float tab. To check the float, look first for damage to it, then shake it by your good ear, listening for gas inside. If you hear it you have to replace the float. Look carefully

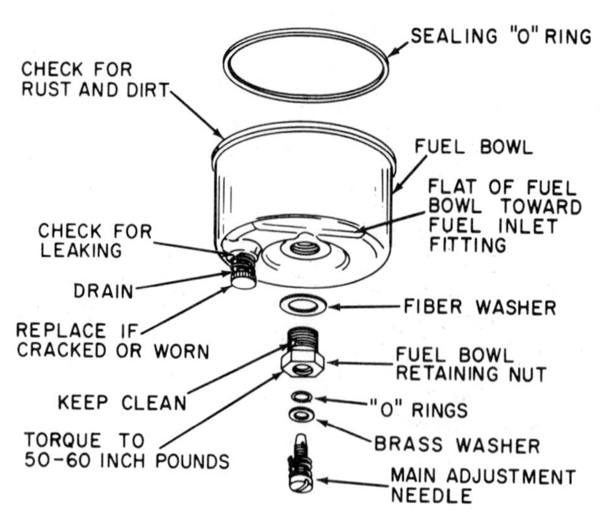

FIGURE 50. Fuel bowl components on Lauson (Tecumseh) type float carburetor.

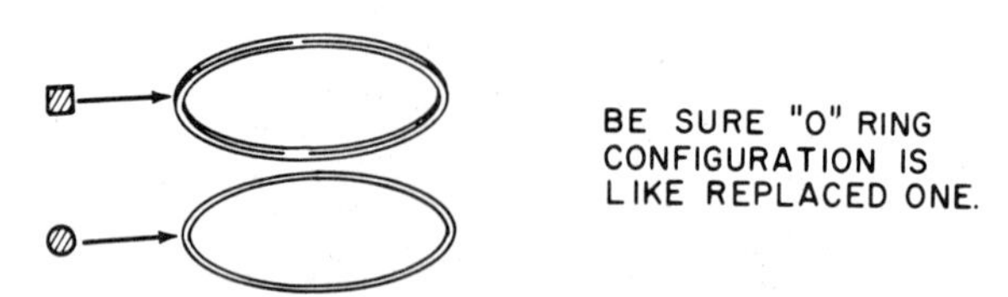

FIGURE 51. O rings on fuel bowl should be replaced with exactly similar new ones. Moisten them with oil.

at the float hinge for signs of wear at the hinge pin surfaces. If wear signs are visible you have to replace the float.

Look carefully at the needle—the tip of it. Some tips are Teflon; others are metal. The tapered tip must not have signs of wear. If it has, replace it. Run your finger across the tip and feel it, as well as look at it in a bright light. Your finger should not feel any roughness or ridge on the tip.

When you replace the float, check its relation to the carburetor surface, using the special plastic tool you find in the carburetor repair kit. It must be parallel. The tab edge of the measurement tool must rest without force on the float, and without any gap between tool and float. If the float is too high or too low, remove the float and bend the tab with a needle-nose pliers. Sometimes you can take a screwdriver and bend the tab without removing the float. But it's no big deal to remove the float, and it is definitely safer.

On these float carburetors the inlet needle and its seat bear a crucial relationship to the correct operation of the engine. They meter

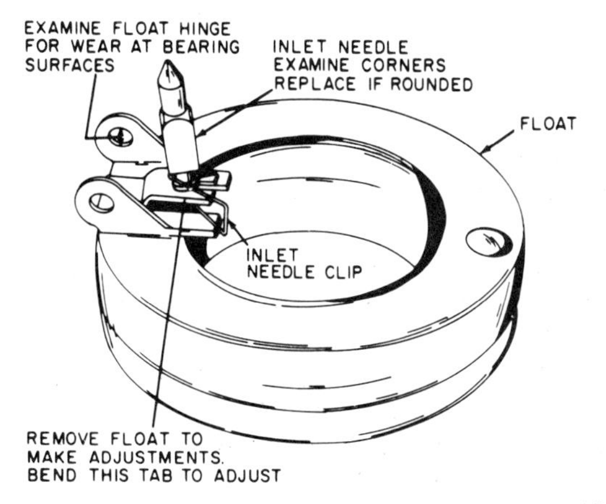

FIGURE 52. Float and needle valve assembly removed for examination.

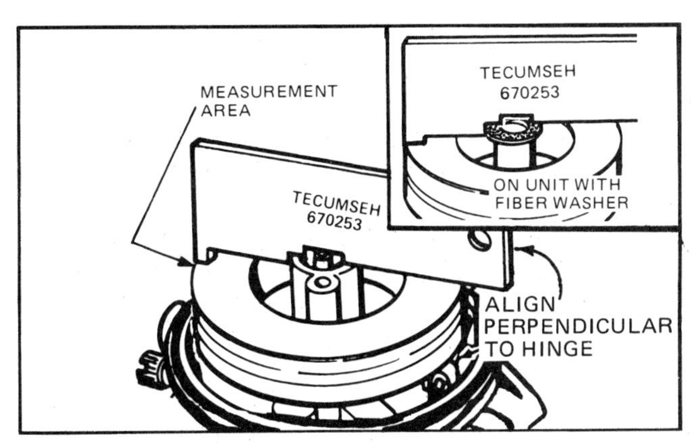

FIGURE 53. Special Tecumseh tool used to measure float height.

the gas with perfect precision, doling it out as the engine requires it. The engine knows what it is using, and if the carburetor cannot cooperate scrupulously the engine cannot run properly. For this excellent reason the inlet needle and its seat and the float are a trinity of relationships that must be precise.

If you get the float back and fill the tank with gas, only to discover that the float sticks, loosen the bowl nut one full turn and turn the bowl ¼ inch in either direction, then return the bowl to its original position. Tighten the bowl nuts.

The inlet needle tip seats into a synthetic rubber fitting in the carburetor body. If you have to replace the seat—when replacing the inlet needle—pull it out with a hook made of wire. Moisten the new seat with oil, and put it in with the smooth side toward the needle, the grooved side down into the carburetor. Use a flat punch about the diameter of the seat and tap it in, carefully.

When installing the inlet needle, get the long, straight end of the clip to face the choke end of the carburetor. If you don't get it right the needle could bind.

Some of these carburetors (Tecumseh) have welch plugs on the side of the carburetor body, just above the idle adjusting screw. They seal the idle fuel chamber that contains the idle and secondary fuel discharge ports. That is why cleanliness is so important. There are several other plugs in the carburetor, and these should not be removed. One such plug is near the inlet seat cavity, on top of the carburetor body. It seals off the idle air bleed—a straight passage drilled into the carburetor throat. Another plug is in the base where the fuel bowl nut seals the idle fuel passage. A third plug is found on the side of the leg that seals the idle fuel passage. What's the point of these plugs if you don't remove them? They are for the benefit of the manufacturer and the remanufacturer, who may need to recalibrate the various passages and venturi.

We have noted that Tecumseh, like Briggs & Stratton, uses float and diaphragm carburetors throughout the various engine lines. Inevitably, there are slight model differences.

You identify carburetors, for purposes of buying a kit, by model number stamped on the carburetor. Tecumseh 89 4F5, a float type, has its own minor differences from other float carburetors by Tecumseh. The differences, though minor, are worth noting.

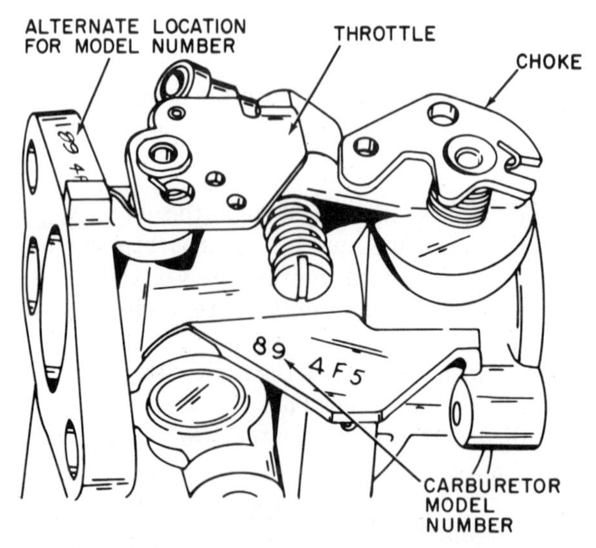

FIGURE 54. Tecumseh carburetor model 89 4F5.

To remove this model carburetor from the engine, remove the carburetor and intake manifold as a complete assembly, and take the carburetor off the manifold later. In the process of disconnection, once you take off the bolts that hold the carburetor and manifold, you then take off the fuel line and grounding wire, the governor linkage (we'll discuss governors later), following no hard-and-fast rules. Use common sense, after you look at all the original connections.

The fuel bowl in this model (and all others in the Tecumseh line) is a key component. The retaining nut contains the transfer passage through which fuel goes to both high-speed and idle systems. It is the larger hole closest to the hex nut (hex or allen—the allen wrench removes the hex nut; it's one of those language puzzles). Look at the small fuel passage in the retaining nut; it must be clean. Also look for rust and dirt in the fuel bowl and clean it all out. Look for leaks at the drain valve, which has a rubber gasket. Simple replacement cures that. The O rings that seal the fuel bowl to the carburetor must be in tip-top shape; that means they need replacement whenever you take the carburetor off. If and when you put on a new O ring, moisten it with oil or gas so the fuel bowl will fit the carburetor properly. Turn the carburetor upside down, put the O ring on the carburetor body, and then put the fuel bowl in the right position. On most Tecumseh engines the fuel bowl flat surface faces the same side of the carburetor as the fuel inlet. But on some outboard motors the bowl drain may be placed differently.

On this carburetor (89 4F5), no inlet seat fitting needs to be removed, but you do examine the inlet needle for wear and replace it if your eye sees wear. Or rub your fingernail across the plastic tip to feel for unevenness. The needle fits into a Viton rubber seat, which may need to be replaced if a visual inspection caused you to suspect it of wear and leaks. The seat comes out easily by bending a hook on the end of a piece of wire and pulling it out.

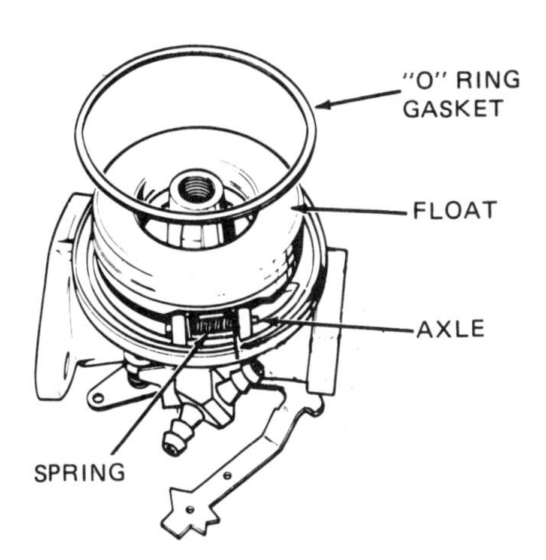

FIGURE 55. Float assembly, O ring gasket on Tecumseh carburetor.

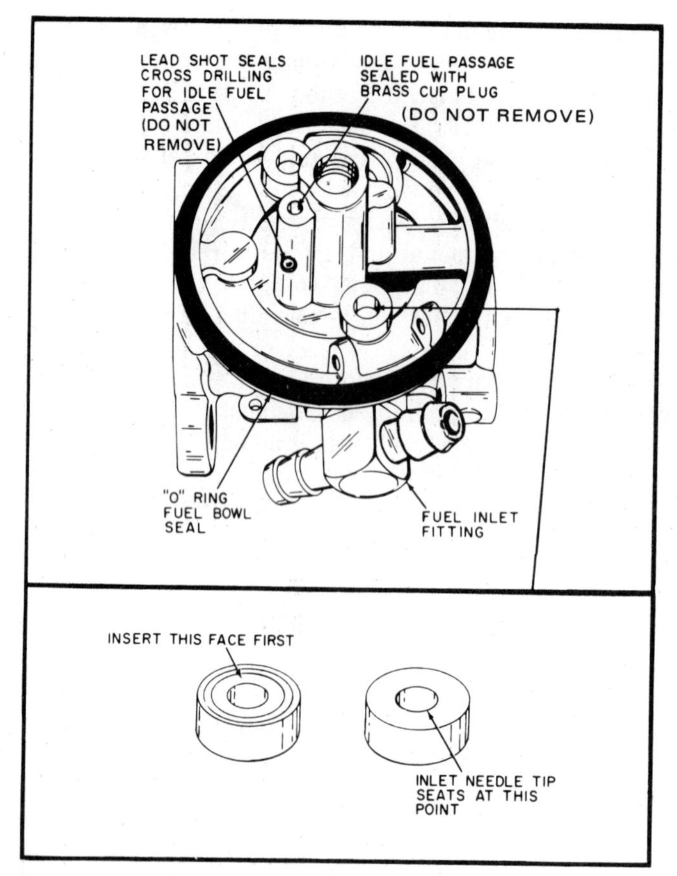

FIGURE 56. Details of fuel inlet system (top), including inlet needle valve seat (bottom).

In this carburetor it may not be mandatory to remove the welch plug. Tecumseh suggests that you can clean out the area covered by the plug with carburetor cleaner or compressed air. It does not make this suggestion in earlier welch plug instructions with other models, but I would strongly suggest it.

A warning: The main nozzle tube, pressed into the body of the carburetor, should not be disturbed. If you push it around its metering characteristics will be changed, ruining the carburetor beyond repair. If the area is dirty, it can be cleaned with cleaning solvent.

Two cup plugs of brass may be noticed. One is in the base where the fuel bowl nut seals the idle fuel passage. Another is near the inlet seat cavity, which seals off the idle air bleed. It's a straight passage drilled into the carburetor throat. There's also a small ball plug on the side of the idle fuel passage.

All these plugs control areas that should be cleaned out with carburetor cleaner, then rinsed with gas or kerosene. You may use dishwasher detergent and warm water, but carburetor cleaner is better.

The Series II is another carburetor similar to the one under discussion. The idle air needle is in a different location and allows an easier adjustment of both air and fuel in the idle section. (The original model permits only the air adjustment.) The Series II has two types of inlet needles and seats.

The first is a resilient Viton seat and hard needle, and the second is a rigid brass seat and resilient Viton-tipped needle. If it is necessary to replace the resilient tip needle, it isn't necessary to replace the seat.

With the Viton seat, use the bent end of a paper clip or wire and make a 3/32-inch hook. Push the hook through the Viton seat hole, then pull the Viton seat out of its brass cup. Replace with a new seat. Don't hook the clip over the brass cup edge.

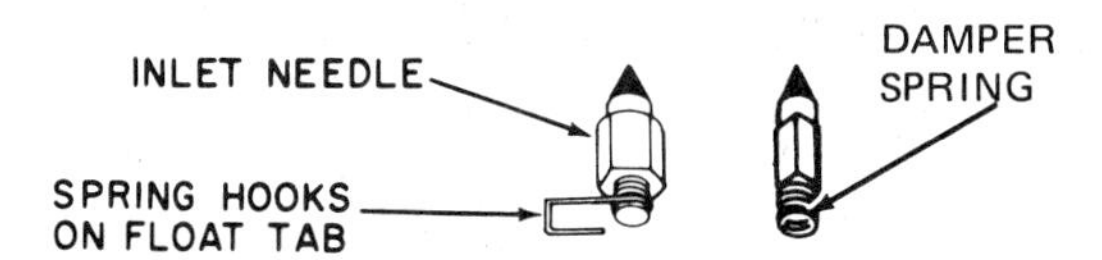

FIGURE 57. Two types of needles on Tecumseh Series II carburetors.

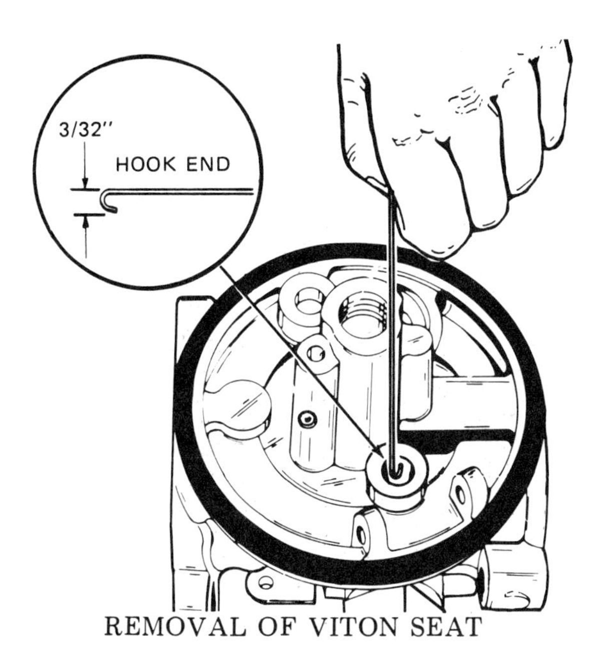

FIGURE 58. Replacing the Viton seat on a Series II carburetor.

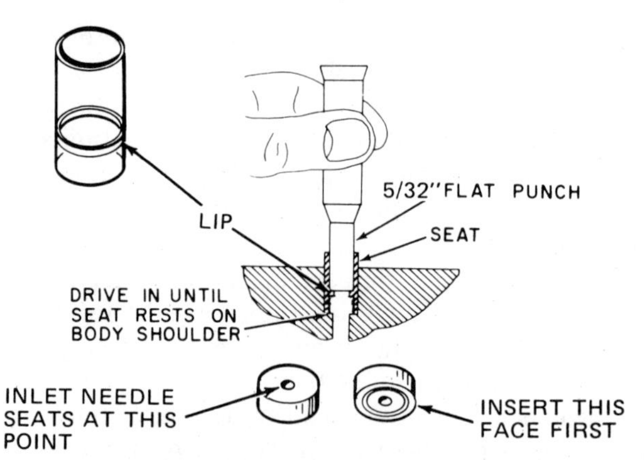

FIGURE 59. Installation of Viton seat.

If necessary to replace the Viton seat, clean out the cavity and place the seat over the cavity with the smooth side up, the grooved side into the brass cup. Use a $^{5}/_{32}$-inch flat punch (socket or whatever), and press the seat into the cavity. The seat must pass over the lip at the base of the brass cup. We have encountered this operation before.

Another Tecumseh carburetor is called the TVS. It has no idle adjusting needle. It has a high-sensitive vane that is attached to the choke shaft. One arm of the vane is attached to the throttle. The choke shutter operates independently of the vane. A sleeve on the choke shaft is adjustable to increase or decrease spring tension, thus regulating the vane and the engine speed. At the base of the vane, adjustments are available for the control to be set to "choke," "high-speed," "idle speed,"

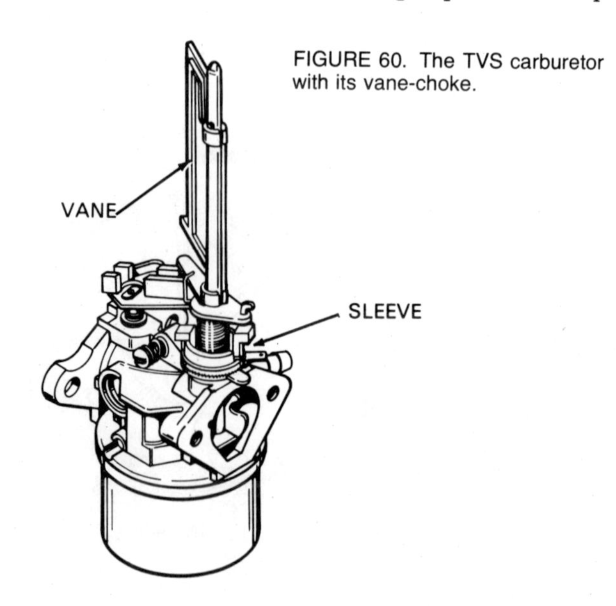

FIGURE 60. The TVS carburetor with its vane-choke.

and "engine cutout." Cutout occurs by positioning a metal tab counterclockwise; the tab contains an ignition cutout lead that short-circuits the ignition ground on the carburetor frame.

Another Tecumseh carburetor is the fixed main jet type, which functions very much like the earlier, fully adjustable types except that the fuel bowl retaining nut has the fixed main jet flowing through it. Service on this carburetor does not differ from service on earlier ones.

The Tillotson MD float carburetor has two idle adjustment screws and a main adjustment needle. When taking this carburetor apart for cleaning and servicing, turn out the main adjustment needle before separating the carburetor halves. Otherwise, the needle will be damaged. Check the clearance between the straight edge and the float. It should be 1/64 inch. If the float level must be adjusted, remove the pivot screw and the float and bend the vertical float lever tang to the correct measurement as in other float adjustments.

The other main carburetor type with diaphragm (the pressure differential) is similar to the Briggs & Stratton Pulsa-Jet in operation, though details are different. The carburetor model is stamped on the flange.

Tecumseh warns about soaking plastic parts in cleaner liquids, but all metal parts can be soaked. Welch plugs should be removed only in carburetors that are extremely dirty. Use carburetor cleaner and rinse with kerosene or gasoline.

These carburetors have idle and high-speed adjusting screws. When you can't get the engine to perform responsively and correctly at both low and high speeds, it is time to track down the two systems. That means disassembly and cleaning of parts, replacing gaskets, O rings,

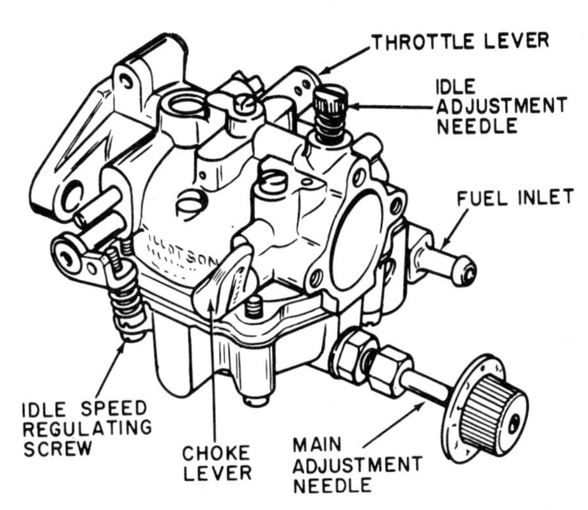

FIGURE 61. Tillotson MD float feed carburetor. Note the two idle adjusters and the main adjustment.

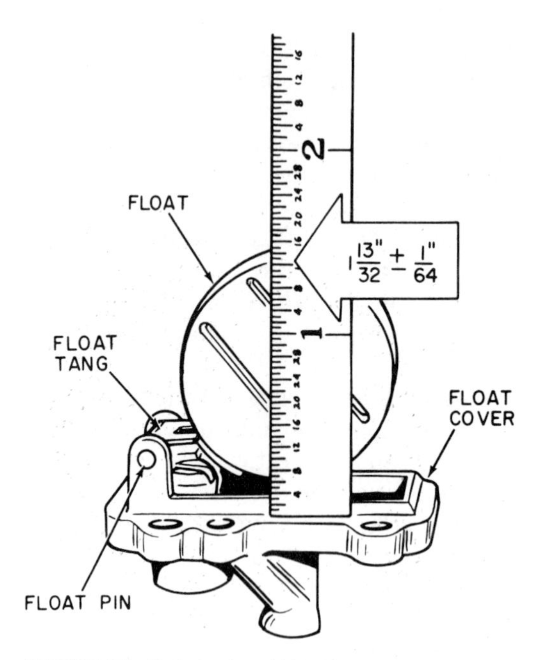

FIGURE 62. Get out the old kitchen ruler for this float adjustment on the Tillotson.

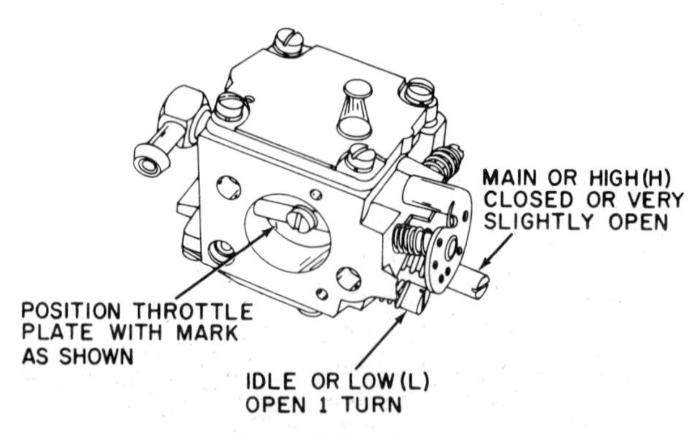

FIGURE 63. The Tillotson HS model has both low- and high-speed adjusters.

and probably the diaphragm, a rather different animal from the Briggs Pulsa-Jet types. When taking out the diaphragm, remove the four retaining screws that hold the diaphragm cover to the body of the carburetor. Examine the diaphragm in a bright light for cracks, pin holes, and stretching. There is only one cure for defects—replacement. When installing the new diaphragm or replacing the old, put the diaphragm rivet head toward the inlet needle valve. It should be noted that

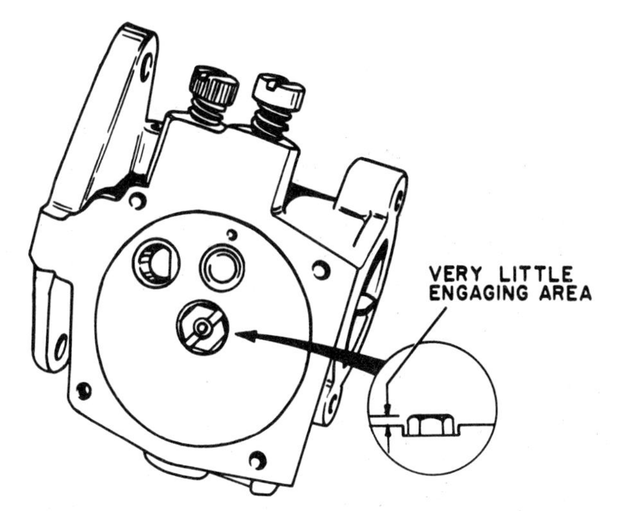

FIGURE 64. There is very little that is engaging about the adjustment or removal of the inlet needle and seat in this Tecumseh. But they admit it.

Tecumseh's diaphragm-type carburetors have a needle inlet valve, a system not present in most such carburetors.

Some of these Tecumseh carburetors require a slotted screwdriver to remove the inlet needle and seat. These are "early types." Later models use a 9/32-inch socket to remove the hex head inlet seat, but Tecumseh notes that there is little space to operate in and you may have to grind off some of the socket. Why couldn't space have been provided? People who design these devices do not overflow with the milk of human kindness, a liquid in short supply, in this field and others too numerous to mention.

The inlet needle has a spring on it, and when you unscrew the needle hold the spring down for the last couple of turns, using finger power to prevent the spring from flying off.

The fuel inlet fitting has a strainer as part of the device. To clean, don't soak in carburetor cleaner because the fitting has a neoprene seat—which should not be removed—so clean with air or use a brush of some sort. If you can't clean the strainer and it appears to be clogged with dirt, you'll have to replace the fitting.

If you deal with a chain saw carburetor, which has an all-position idle system (these things have to work upside down), note that the idle circuit is a lulu. It runs through the carburetor body—not merely a sector of it—through gaskets, offsets, and then to bottom parts of the engine crankcase, where it atomizes the excess fumes and fuel.

Another carburetor in this series is the Tillotson HC, which has three adjustment screws—an idle regulator, idle adjustment, and main adjustment needle screw. This carburetor differs in details from others.

It is more complex, containing a metering gasket, a metering diaphragm, and a pump gasket and pump diaphragm. It is used in chain saws (two-cycle engine). It must be free of sawdust before disassembly. In view of its complexity, try everything else *before* taking it apart. It is a good thing that George Washington didn't have to employ a chain saw in the cherry tree episode, especially one with a Tillotson carburetor. When I say try everything else, I mean use a quick-flush carburetor cleaner on it, following directions carefully so as not to ruin the plastic parts and the neoprene or rubber components. Then try new adjustments—they go astray more readily than rabbits.

If these tactics fail—and I do not wish to be optimistic about them—you face a major disassembly job, rivaling a Holley carburetor on a new car. This teaches patience to an unprecedented degree.

Disassembly goes like this:

Remove the pump diaphragm cover screws and cover, pump gasket and pump diaphragm, filtering screen, main diaphragm cover screws and cover, main diaphragm and gasket by sliding the diaphragm toward the adjustment screws 1/16 inch, and pull it up so as to unhook it from the control lever. Next, remove the fulcrum pin screw, pin, control lever and spring, and the inlet needle.

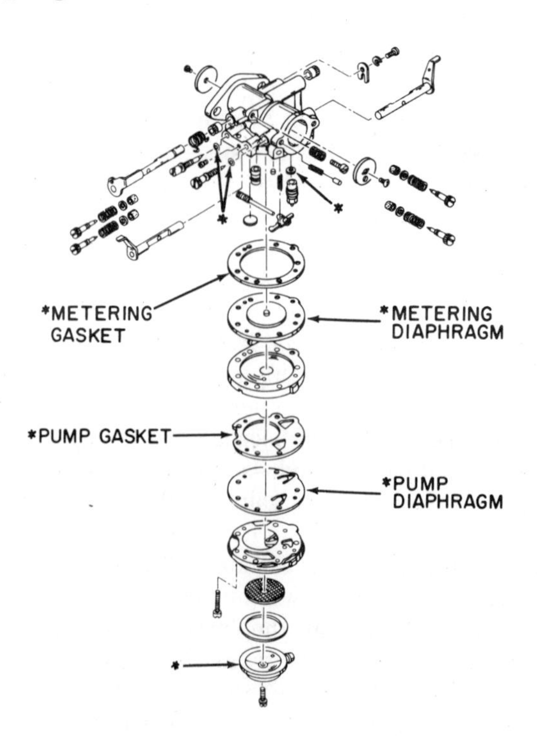

FIGURE 65. The Tillotson HC, found on chain saws and in other equipment.

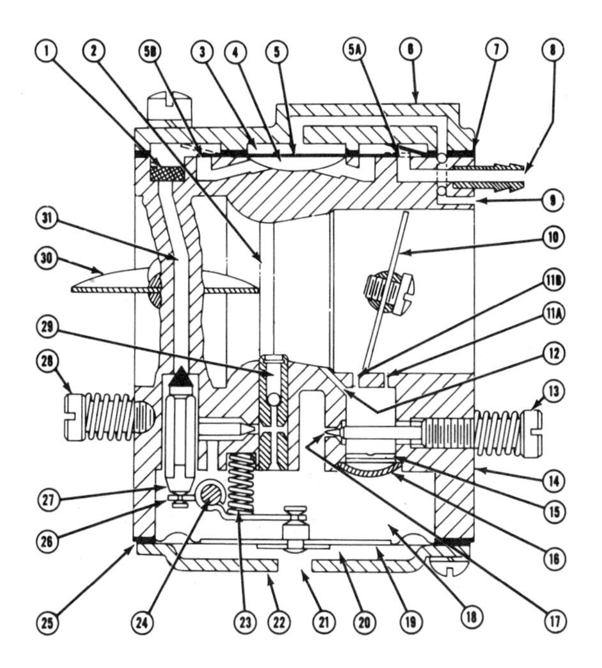

1. Filtering Screen
2. Venturi
3. Pulse Chamber
4. Fuel Chamber
5. Fuel Pump Diaphragm
5A. Diaphragm Pump Inlet Valve
5B. Diaphragm Pump Outlet Valve
6. Fuel Pump Body
7. Fuel Pump Gasket
8. Fuel Inlet
9. Impulse Channel
10. Throttle Shutter
11A. Primary Idle Discharge Port
11B. Secondary Idle Discharge Port
12. Air Bleed Passage
13. Idle Fuel Adjustment Screw
14. Body
15. Body Channel Reducer
16. Welch Plug
17. Idle Fuel Adjustment Orifice
18. Metering Chamber
19. Diaphragm
20. Atmospheric Chamber
21. Atmospheric Vent
22. Diaphragm Cover
23. Inlet Tension Spring
24. Fulcrum Pin
25. Diaphragm Gasket
26. Inlet Control Lever
27. Inlet Needle
28. Main Fuel Adjustment Screw
29. Main Nozzle Discharge Port and Check Ball Assembly
30. Choke Shutter
31. Fuel Inlet Supply Channel

FIGURE 66. This Tillotson HS carburetor follows HC patterns of disassembly and repair. They are much alike. The formidable list of components doesn't mean you have to disassemble them all.

Soak all metal parts and rinse.

In reassembly take care that the inlet control lever and spring go back correctly. The spring rests in the well of the metering body. It touches the dimple of the inlet control lever. The control lever should be flush with the floor of the diaphragm chamber. If the diaphragm end of the control lever is low, pry it up. If the lever is high, depress the diaphragm and push on the needle for adjustment. There are some models in which the inlet control lever is hooked to both the inlet needle at one end and the metering diaphragm at the other. Old carburetor hands are used to this kind of arcane complexity; it comes with the territory.

Next, you deal with the main nozzle and check ball. Remove the main adjustment screw. Drive the check ball assembly out of the body and into the venturi, using a punch and hammer (it doesn't matter what happens to the assembly—you're replacing it with a new one—but don't bang the walls surrounding it). Install the new one from the metering chamber side, flush with the metering chamber surface. Use great care when driving the new one in. The driving surface should cover what you are driving, and the force should be just enough.

The channel reducer (it takes up space) is a loosely floating measuring device, in the idle fuel chamber. It has one correct and one incorrect way to be installed; so, notice where the mark is at its center—it faces up. However, the chamber is concealed by the welch plug, which you should not disturb unless all else fails—that is, if you can't clean out the carburetor well enough to restore original engine performance.

The throttle shutter, screwed into the throttle shaft and lever assembly, fits in only one way. It is not round. The small mark on the shutter should be positioned as shown. The mark can be located either above and to the right of the shaft or below and to the left of it. You remove the shaft clip, relieve the tension of the retainer spring, and lift the shaft from the body.

If the choke binds and requires service of any sort, it slides out after you have flattened the raised tab of the choke shutter. As the choke shaft lifts from the body, a friction ball and spring will pop out if you don't hold them back.

In reassembly, slide the choke shutter back into the shaft slot until the two raised stops hit the choke shaft. Then, with a long punch or screwdriver, strike the choke shutter on the tab near the slot. This tab secures the shutter.

Tecumseh also uses carburetors of this type—diaphragm—by Carter, best known for automobile carburetors. Carter makes simple, easy-to-service auto carburetors, and the ND series for Tecumseh does not disappoint in these matters. The ND carburetors have a fuel pump with housing on top. They are unexceptional in other matters.

KOHLER CARBURETORS

Kohler engines, like Tecumseh, Briggs & Stratton, and other small, one-cylinder engines, are found on various powered devices. They are found with simple rope starters (as on push-type lawn mowers), or with starting motors, batteries, charging systems, and electronic ignition (on larger devices).

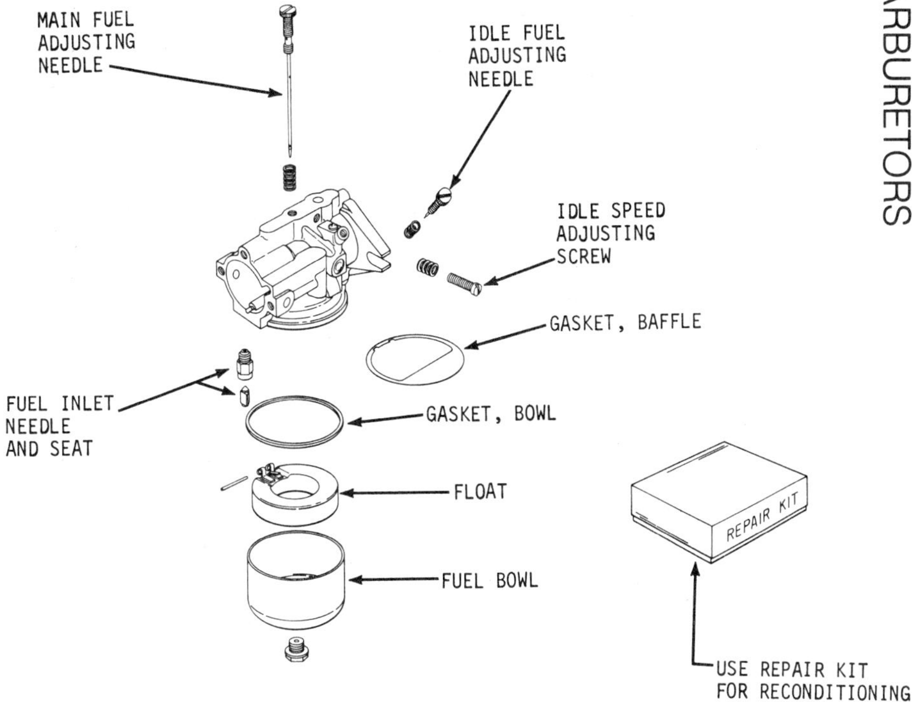

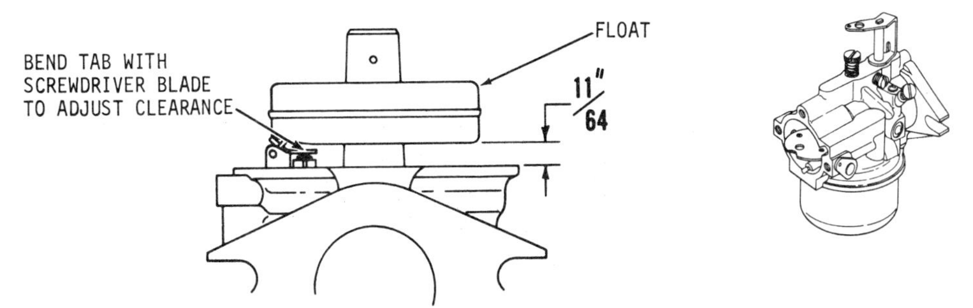

FIGURE 67. The Kohler side draft carburetor.

Two types of carburetors may be found, the side draft and the up draft. This simply refers to the direction from which air comes into the carburetor throat, with the necessary adjustments to parts placement.

To take the side draft type apart, once you take it off the engine, remove the bowl nut and gasket, then the bowl. Some of them have a bowl drain, which consists of a drain spring, spacer, plug, and gasket. Next, remove the float, needle, and needle seat, checking the float for leaks and wear on the float lip and the float pinholes (what the pin fits through). If the float wobbles the least bit on the pin, it means you must replace the float. Many a carburetor has come to grief on this elementary matter, and many are the technicians who can't figure out why the carburetor (thus the engine) misbehaves, when the reason is this simple wearing away of the pinholes by the pin. So, before pulling the pin out, push the float around on the pin, looking for looseness and wobble. There should be none. These caveats apply to every carburetor, from the two-cycle engines we've talked about to the Quadrajet on your Chevrolet Blazer.

Remove the adjusting needles, but do not take out the choke and

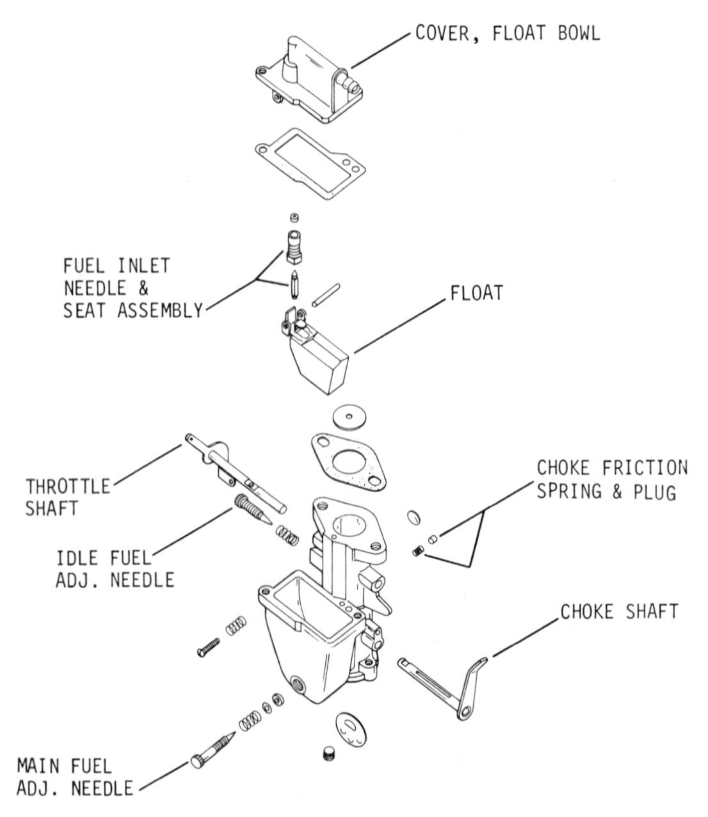

FIGURE 68. Kohler's up draft carburetor.

throttle plates and shafts. If these things go bad in a Kohler, you have to buy a new carburetor.

The up draft disassembly requires similar removals—bowl cover and gasket, float pin and float, needle and seat. These are simple carburetors in a field that long ago forgot the old saw about simpler being better. It is, it was, but it will never be again.

Kohler small engines have a fuel pump that is fairly sophisticated, but not terribly complicated, in all their engines except model K91. It operates off a cam on the camshaft, just like the fuel pump on most cars. In fact, it is an automobile-type pump. The difference is that unlike modern auto fuel pumps, this one can be disassembled and repaired, whereas your typical auto pump can only be replaced. Fortunately, the progress toward the disposable car has not yet reached the small engine sector of the economy. However, before we hail Kohler for its progress *back* to the repairable fuel pump, note that one Kohler pump cannot be repaired, only replaced. It's one that is adapted for the K91 engine that doesn't usually take the regular pump, but in some applications it is necessary to graft a pump on, after all. In these cases an adapter is needed, plus other stray pieces of hardware. It's a rare application.

Repair kits are available for all other Kohler fuel pumps.

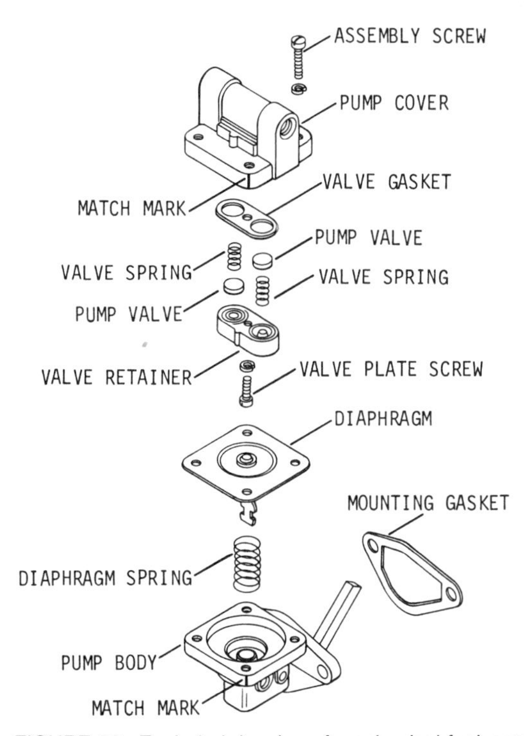

FIGURE 69. Exploded drawing of mechanical fuel pump.

CONDITION	POSSIBLE CAUSE/PROBABLE REMEDY
A. Black, sooty exhaust smoke, engine sluggish.	A. Mixture too rich - readjust main fuel needle.
B. Engine misses and backfires at high speed.	B. Mixture too lean - readjust main fuel needle.
C. Engine starts, sputters and dies under cold weather starting.	C. Mixture too lean - turn main fuel adjustment 1/4 turn counterclockwise.
D. Engine runs rough or stalls at idle speed.	D. Idle speed too low or improper idle adjustment - readjust speed then idle fuel needle if needed.

FIGURE 70. This brief troubleshooting chart (by Kohler) applies to all carburetors everywhere.

CARBURETOR REMOTE CONTROLS

On Briggs & Stratton engines one finds three types of remote controls: governor, throttle, and Choke-A-Matic. Governor controls on small engines have no counterparts on large ones. They accelerate or retard the engine depending on what it's doing. If the going gets tough, the engine gets going, thanks to the governor, and the converse is likewise true; that is, if the going gets simpler, the engine slows down. It replaces your foot on the accelerator, and it is less problematic than your foot. (No offense.)

The way the governor control works is that it changes spring tension on the governor and thus regulates engine speed. We will see much more about governors, below.

The remote throttle control resembles the auto accelerator in that it allows the operator to control speed up to a point. Then, unlike the acclerator pedal, the governor takes over and, when the full governed speed is reached, the governor prevents high, possibly damaging speed. Cars should be so considerate. The Choke-A-Matic remote control controls both the choke and the stop switch. If the remote control fails to

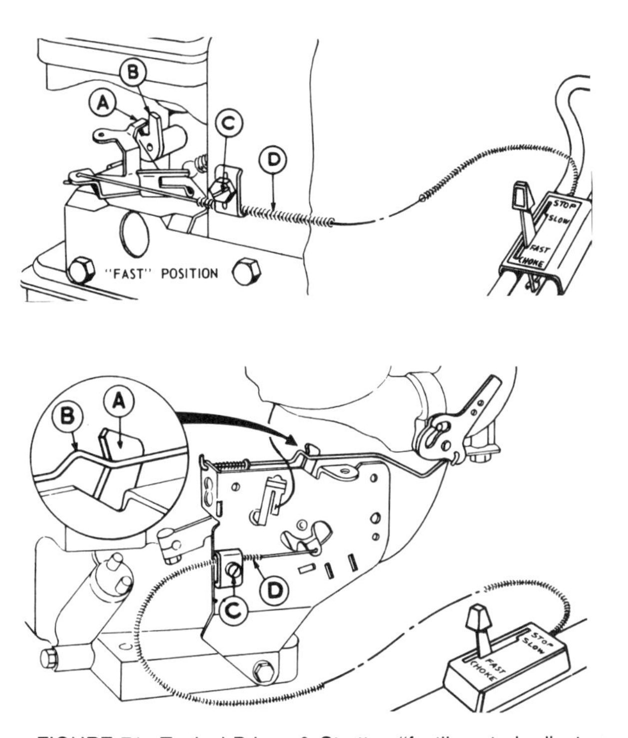

FIGURE 71. Typical Briggs & Stratton "fast" control adjusters.

close the choke completely for starting and fails to open it completely once the engine is running properly, and if it fails to turn the engine off, the control needs adjusting. It won't fail in all these situations simultaneously, but it has to do all three of them in order.

We have already noted the procedure for the basic "fast" adjustment on these Briggs engines. It varies little from model to model.

Choke valve remote control operations also follow similar patterns from model to model. For adjustment, move the remote control choke lever to either "start" or "choke" position. The choke valve in the carburetor throat must be closed completely. Then move the remote

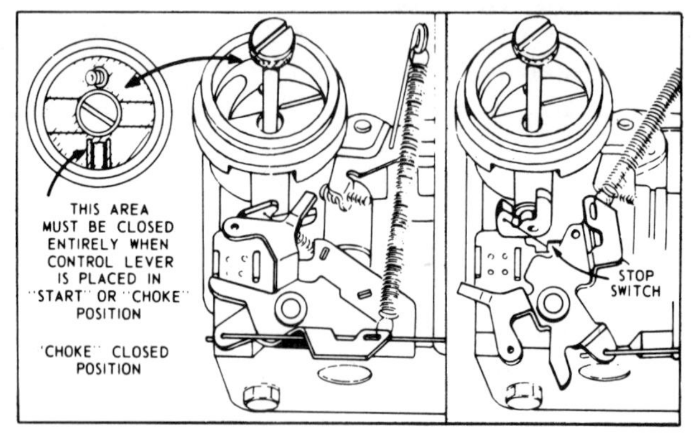

FIGURE 72. Typical choke controls on Briggs & Stratton engines.

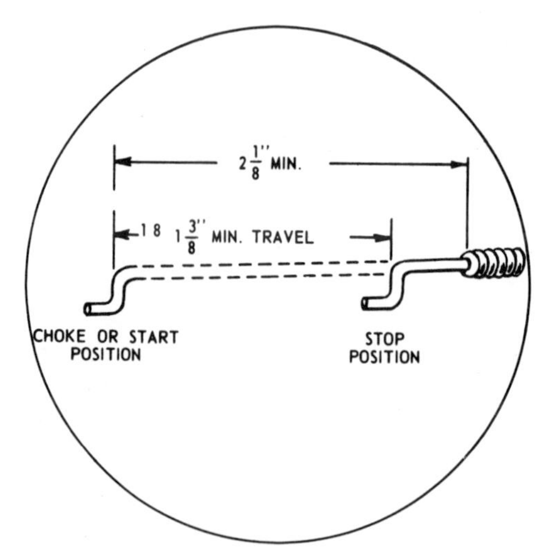

FIGURE 73. Control wire travel on choke adjustments.

control lever to the "stop" position. The control on the carburetor should hit the stop switch blade. Remote control wire travel has to be a minimum of 1⅜ inches for proper "choke" and "stop" positions.

Dial controls on the carburetor need be adjusted only after a disassembly that includes the blower housing. The controls involve only a short length of wire from the control screw over to the carburetor bracket; place the dial control knob in the "start" position, loosen the screw (at "A" in illustration), and move the choke lever "C" to the full throttle position. Allow a ⅛-inch gap between the lever and the bracket (see illustration), and tighten the screw.

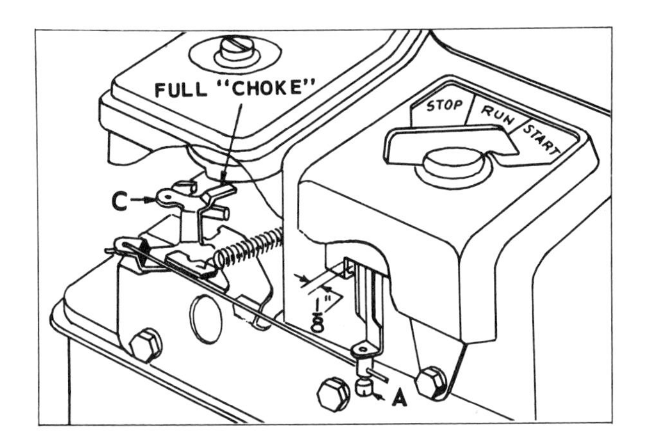

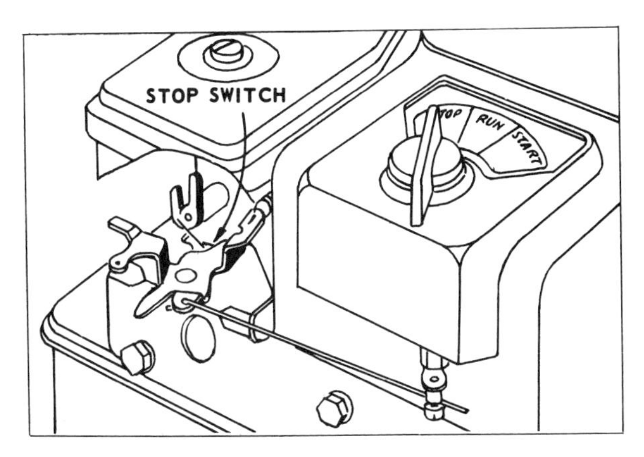

FIGURE 74. Typical dial control settings on Briggs & Stratton engines.

GOVERNORS

Governor springs on Briggs engine models 92500 and 92900 (common lawn mower engines) have double-end loops. When removing and installing these springs do not use a needle-nose pliers, because of the possibility of bending the loops. Follow the procedures illustrated.

The purpose of the governor is to maintain engine speed within fixed limits, even though the load may vary from time to time. There are two main types on small engines—air vane and mechanical types—with a variety of subtypes. As engine load increases, engine speed is slowed and vice versa.

On the air vane governor, which is a metal plate alongside the flywheel, the spring attached to it tends to open the throttle. That increases engine speed. Air pressure against the air vane tends to close the throttle, which reduces the flow of fuel to the engine, decreasing engine speed. As speed decreases, the amount of air directed against the air vane decreases. The governor spring tension then takes over again and turns the air vane toward the flywheel fan, repeating the cycle of engine speed increase and decrease as the load dictates.

This almost foolproof system is maintained by the simplest of means with the air vane type; mechanical types add components, hence complexities.

The typical lawn mower engine uses air vane types. The only things that get out of whack are the spring and the links. Springs lose tension, so it is well to replace the governor spring every two or three seasons, or whenever the system stops working—when tall grass stalls out the engine, for example. You can't tell about spring tension merely by looking at it, so if there is any doubt, replace it.

The mechanical governor also has a spring, but it has gears, weights, and links. The spring pulls the throttle open; the counterweights pull it closed. That's the way the mechanical system works, as load on the engine varies with the task.

The governor is supposed to keep the engine from turning the cutting blade tip of the lawn mower in excess of 19,000 feet per second. That's dangerous. You need a tachometer to know exactly how many revolutions per minute the engine is turning to know the speed of the blade. Revolutions per minute translate into blade tip travel, based also on blade length—the longer the blade the slower the engine needs to run to obtain the 19,000-feet-per-second maximum. Blade lengths vary commonly from 18 inches to 23 inches (some riding mowers go up to 26 inches). At 18 inches, the maximum engine r.p.m. is 4,032; at 19 inches, it's 3,820 r.p.m.; at 20 inches, it's 3,629; at 21 inches, it's 3,456. The 26-inch blade has a 2,791 maximum rotational r.p.m. As noted, you would need a tachometer to do this precisely, but anyone can hear when

CORRECT POSITION OF SPRING

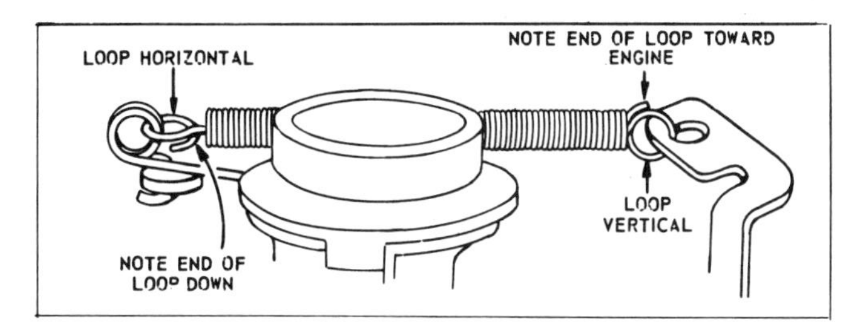

REMOVING SPRING

(1) REMOVE SPRING FROM CONTROL LEVER

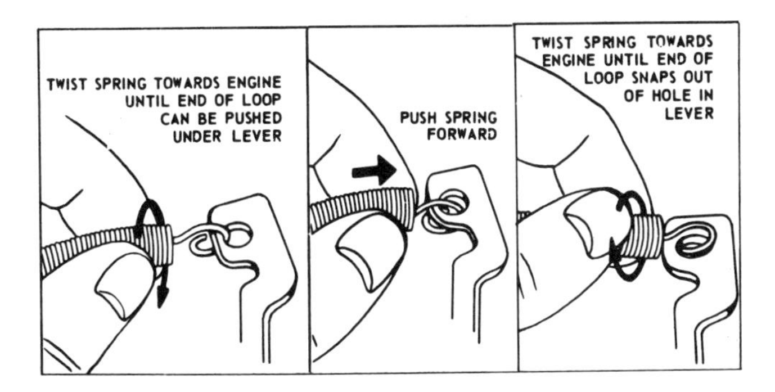

(2) REMOVE SPRING FROM EYELET IN LINK

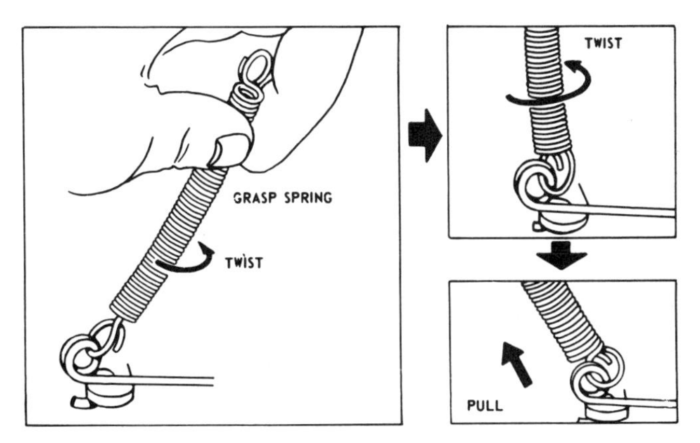

FIGURE 75. Governor controls involve the spring and its loops. These loops must be connected (and disconnected) as shown.

INSTALLING SPRING

① ASSEMBLE SPRING TO LINK EYELET

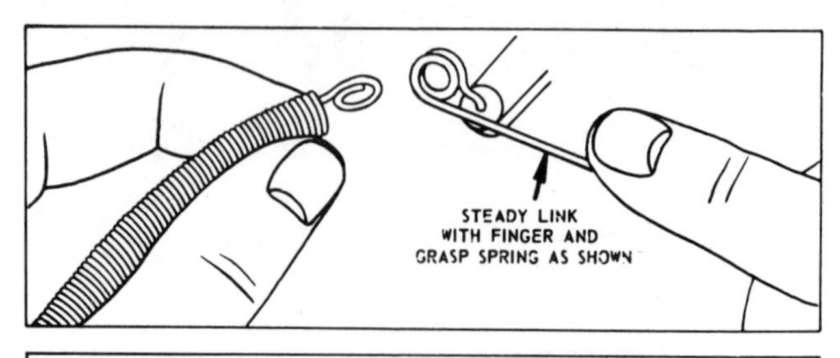

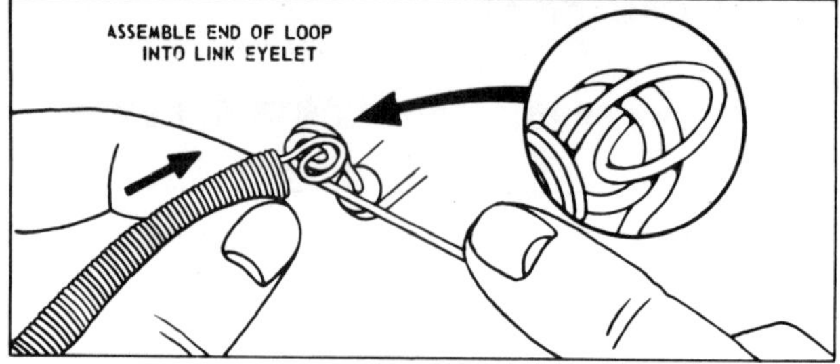

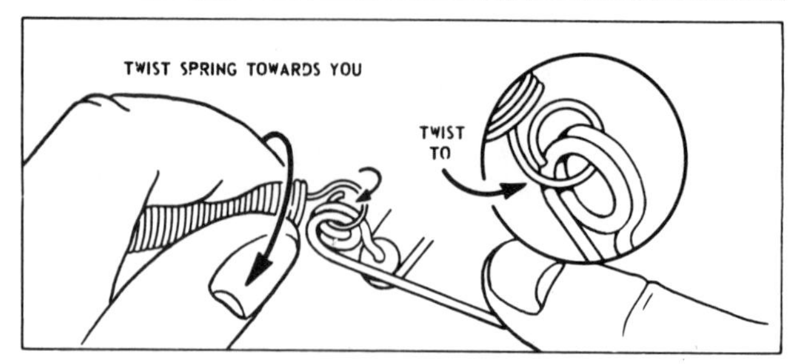

② ASSEMBLE SPRING TO CONTROL LEVER

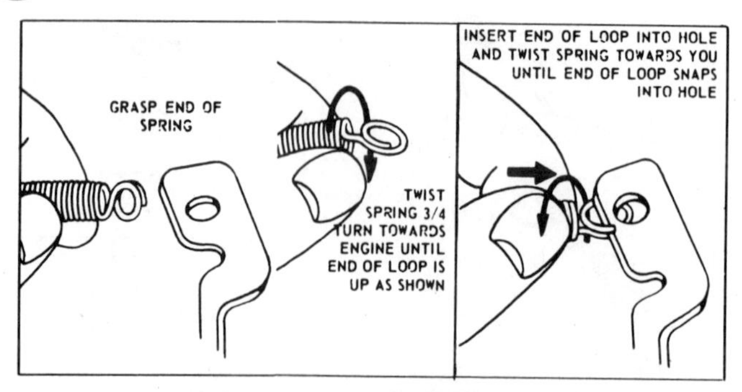

FIGURE 75. Governor controls involve the spring and its loops. These loops must be connected (and disconnected) as shown.

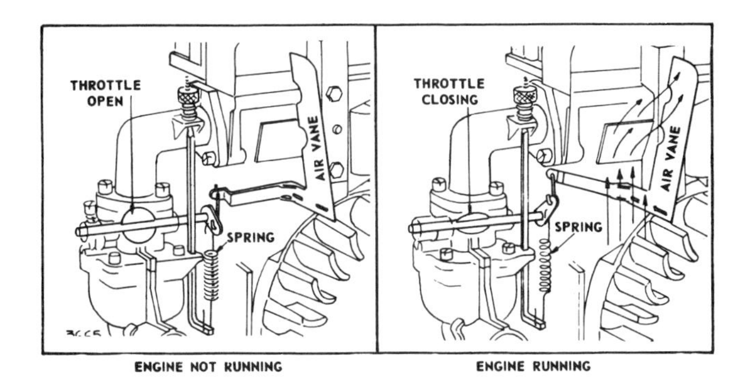

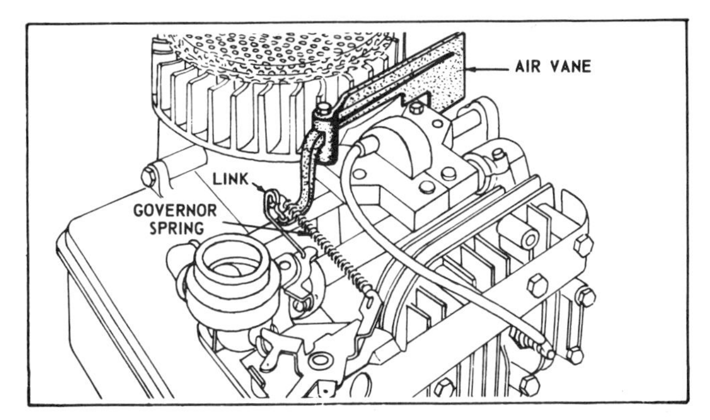

FIGURE 76. An air vane governor shown in action (engine running), and with engine stopped (above). Below, shown up close and in clearer detail. Briggs & Stratton.

a small engine is going too fast—the slower the better, consistent with the job that needs to be done, and no grass cutting requires excessive engine speed.

If the mechanical governor stops doing its duty, you have some disassembly on your hands—unlike the air vane type, which consists of just what you can see alongside the flywheel fan, the spring and link. The mechanical governor has a housing with two mounting screws, a cup, gasket, crank, lever, pin, and gear. The governor gear is driven off a gear on the camshaft gear section (these little engines don't have true camshafts, but cam gears).

Mechanical governors, which are found in larger engines, on Briggs models N, 6, and 8, and on 6B, 8B, 60000, 80000, and 140000, have larger jobs to do, hence their greater complexity. Before tackling a disassembly, consider adjustment. The lever and crank have no adjustments, but by inserting the link or spring in different holes of the governor and throttle levers you change the governor's response and so alter engine response. The closer to the pivot end of the lever, the less

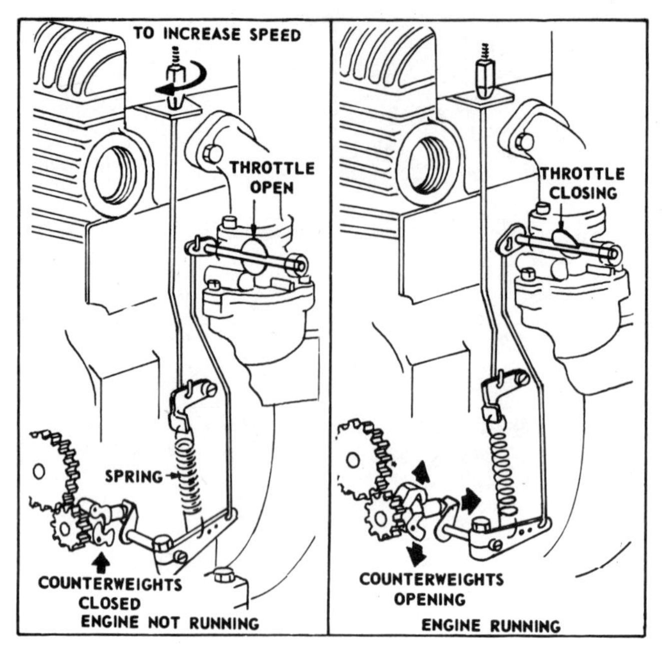

FIGURE 77. Mechanical governor shown in cutaway, engine not running; engine running.

difference there is between a load and no-load engine speed. (Translation: the governor effect is minimized.) An engine will, of course, run perfectly well without a governor, but it's like a car driven in hilly country without your pushing on the gas when it goes uphill. If your small engine is engaged exclusively in unruffling tasks, you can forget everything about its governor. The exception is with a mechanical governor and its gear and housing. The housing can leak and the gear can wear sufficiently to cause damage to the crankshaft. These are remote possibilities.

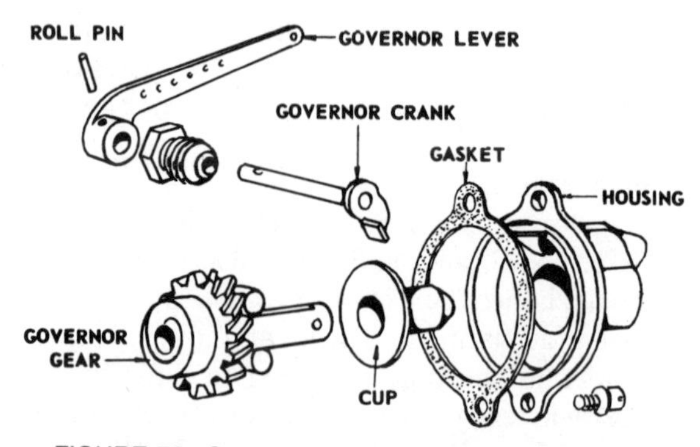

FIGURE 78. Governor housing and gear assembly.

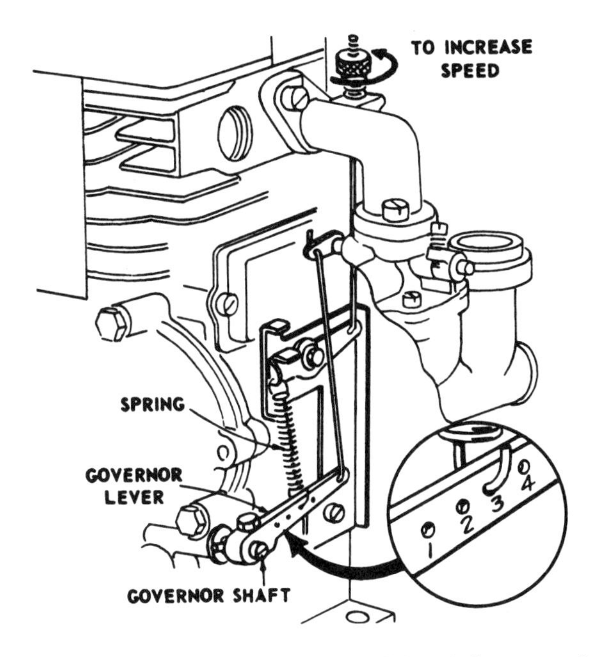

FIGURE 79. Governor adjustments on Briggs & Stratton engines.

In disassembly you have few parts to worry about. Once you pull the cup off the governor gear, the gear will slide off the shaft. The governor crank comes apart when you drive out the roll pin at the end of the governor lever. Then you can unscrew the bushing. In replacement, a new gasket must be used on the housing.

There are other models using these mechanical governors, including Briggs engines with model numbers 6B, 8B, 60000, 80000, and 140000, as noted. The governors are very similar. Adjustment on the governors of these, and other, models goes like this: Loosen the screw

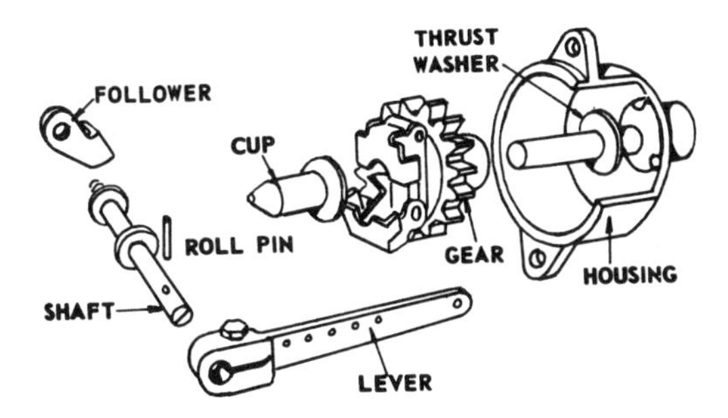

FIGURE 80. Components of mechanical governor found in various models.

that holds the governor lever to the governor shaft. With the throttle in the high-speed position, turn the governor shaft counterclockwise as far as it will go, using a screwdriver. Tighten the screw that holds the governor lever to the governor shaft. Then manually move the governor linkage to determine that it moves freely, as it must.

In some Briggs models it is necessary to do a lot of digging to remove the governor, including removing the engine base. These are larger engines.

TECUMSEH GOVERNORS

Most Tecumseh governors, and all of recent vintage, are of the mechanical type if the engines are four-cycle. These include 3-h.p. to 10-h.p. engines. You find both mechanical and air vane governors on the two-cycle Tecumseh engines.

To adjust air vane governors (two-cycle engines), loosen the self-locking nut that holds the governor spring bracket to the engine crankcase. Adjust the spring bracket to increase or decrease spring tension. Increasing tension increases engine speed; decreasing spring tension decreases engine speed. The spring bracket should not be closer than $^1/_{16}$ inch to the crankcase at point "A" (illustration). Tighten the self-locking nut and start the engine. If you have access to a tachometer you can check engine speed, which should not be over 4,700 revolutions per minute, preferably slower.

On two-cycle Tecumseh engines there are two types of mechanical governors, the power takeoff end and the throttle control types that operate off the flywheel.

Tecumseh engines have a third type of governor, the idle governor to control idle speed exclusively. This arrangement consists of an idle governor cam, a spring, throttle control lever, and an adjusting screw. The governor controls throttle shutter movement at idle speed only. As air rushes through the carburetor air horn it strikes the throttle shutter nearby, closing it. Spring tension in the idle governor tends to hold the throttle open slightly. As the closed throttle reduces air velocity, the

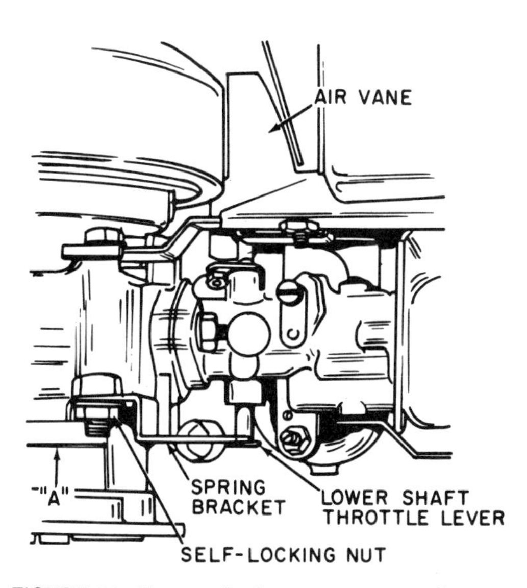

FIGURE 81. Tecumseh air vane governor adjustment.

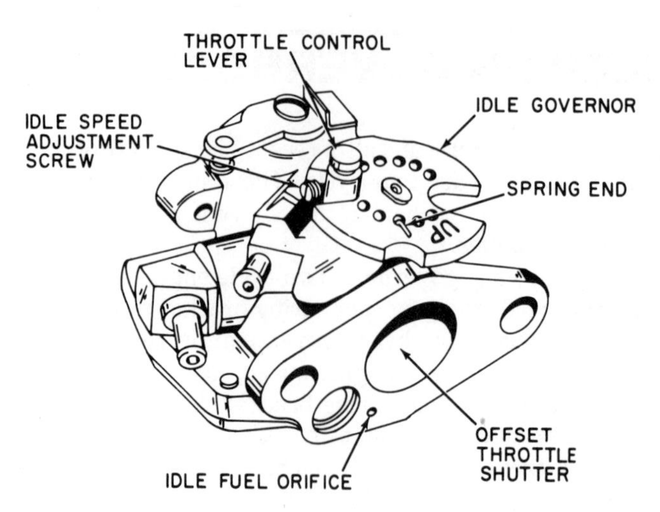

FIGURE 82. Idle governor in two-cycle Tecumseh engine.

spring tension on the throttle shaft overcomes the reduced pressure of the air and reopens the throttle. Binding from dirt and wear is about all that can trouble the operation of this device. If its operation is troubling—and faulty idle is usually caused by either ignition or fuel problems or a combination of them—the order of disassembly is to note first the position of the throttle shutter and the reference marks on it, which must be positioned in one way. Remove the shutter fastener and it will come out of the air horn. Next, note the hole to which the spring is attached in the disk-shaped throttle lever. The spring should go back into that hole. Remove the retainer clip and pull out the throttle shaft. Examine all the parts and replace any that seem worn, and assemble in reverse order to disassembly.

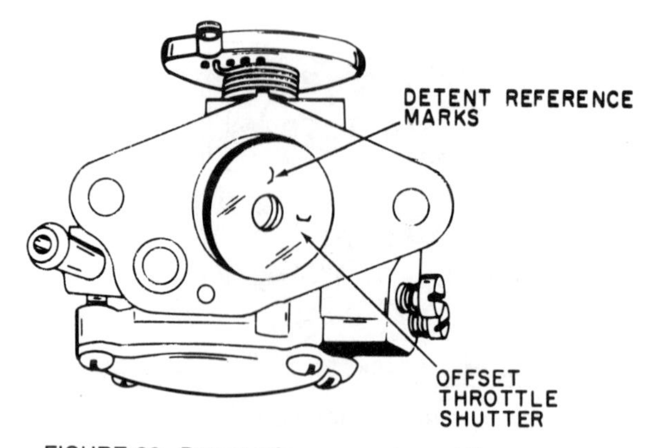

FIGURE 83. Detent reference marks on idle governor.

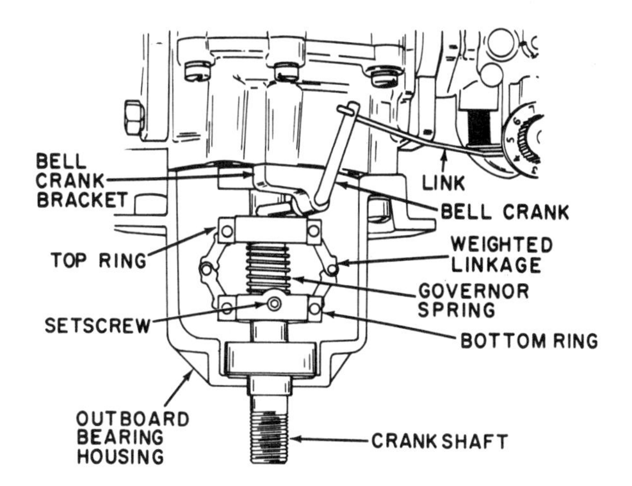

FIGURE 84. A power takeoff governor by Tecumseh.

Before taking anything apart one should always rule out possibilities elsewhere that could be causing the trouble. The fuel and ignition systems should always be checked first. If they are in good working order and the idle still isn't stable, the first thing to look at in the governor is spring tension, Too much tension will cause the engine to surge; too little tension will slow the engine idle.

Governors in the power takeoff end of Tecumseh two-cycle engines consist of two rings, a weighted linkage, and a spring. The lower ring is secured to the crankshaft, so the governor assembly rotates with the crankshaft. As speed increases, the weighted linkages are thrown outward by centrifugal force. The top ring moves toward the bottom ring. The governor bell crank contacting the top ring follows it, permitting the throttle to close, thus slowing engine speed. As the centrifugal force decreases, the governor spring moves the top ring upward, overcoming the throttle spring to open the throttle and thus maintain constant engine speed.

This seems like an intolerably complicated device, but it is simplicity itself. To adjust, when engine speed doesn't seem right and all else checks out, loosen the setscrew and adjust the position of the governor on the crankshaft. Move the governor assembly toward the crankcase to increase engine speed. To decrease speed, move the governor assembly away from the crankcase. Then tighten the setscrew.

To get at the power takeoff governor, remove the screws that hold the outboard bearing housing. Loosen the setscrew that holds the governor assembly to the crankshaft. Adjustment is now possible.

The flywheel governors on Tecumseh two-cycle engines operate through a system of links and link springs. As speed increases, the links are thrown outward, compressing the link springs. Links apply thrust

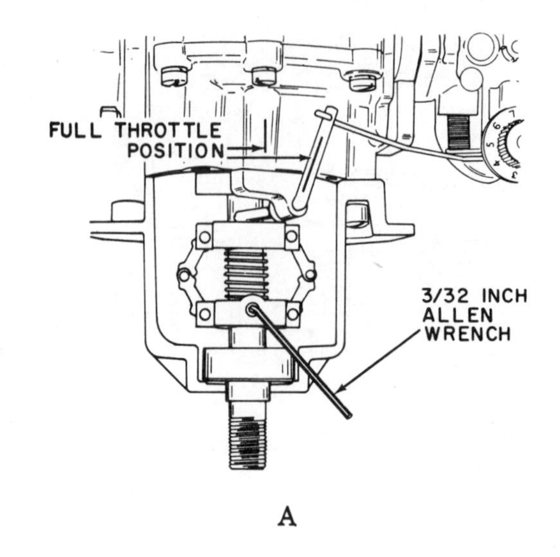

A

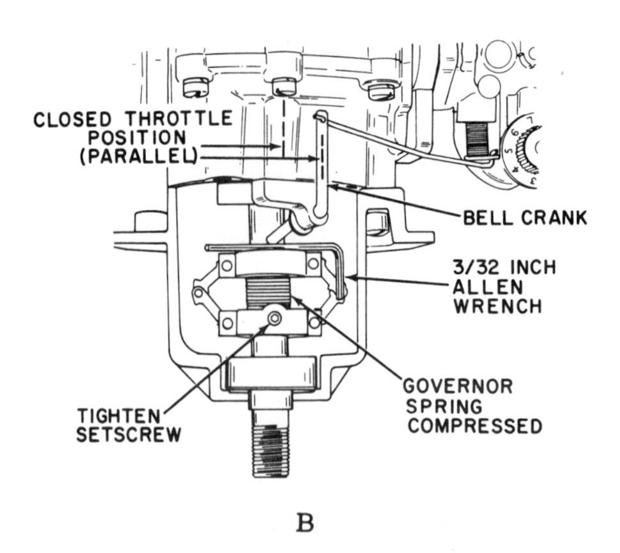

B

FIGURE 85. Adjustments on the power takeoff governor.

against the slide ring, moving it upward and compressing the governor spring. As the slide ring moves away from the thrust block of the bell crank assembly, the throttle spring causes the thrust block to close the throttle slightly. Engine speed decreases as the throttle closes. Force on the slide ring decreases, causing it to move down. This downward movement overcomes the force of the throttle spring and opens the throttle, thus increasing engine speed. The ebb and flow of these centrifugal forces keeps engine speed in balance.

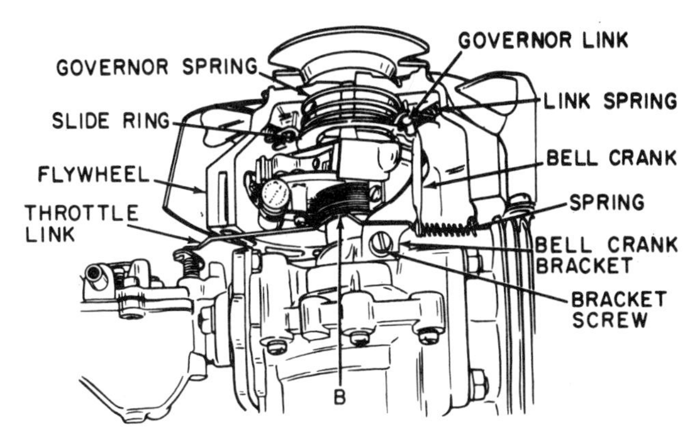

FIGURE 86. Flywheel governor adjusting points.

To adjust these flywheel governors, loosen the bracket screw and slide the governor bell crank assembly toward or away from the flywheel. Moving the governor bell toward the flywheel increases engine speed; moving it away from the flywheel decreases speed. Tighten the screw.

You can make minor speed adjustments by bending the throttle link at point "B" in Figure 86. Bending the link increases speed. Remove the link before bending it; otherwise, you risk distorting the bell crank or throttle lever.

If the carburetor has a throttle lever with more than one hole, place the throttle link in the hole nearest the lever pivot point. When the throttle is in the open position, the thrust block must clear the breaker point box by at least $^{1}/_{32}$ inch but not more than $^{1}/_{16}$ inch.

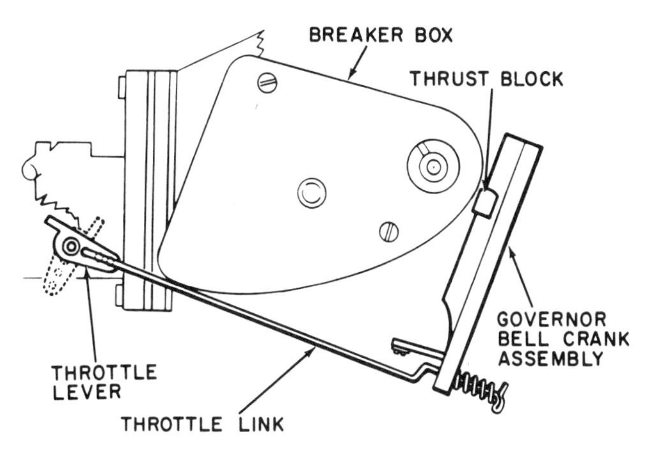

FIGURE 87. Checking flywheel governor adjustment.

One other governor found on some two-cycle Tecumseh engines is a safety net kind of device. It is called a two-cycle ignition governor (throttle control type), that shorts out the ignition system when the engine reaches a maximum r.p.m. range from 4,400 to 5,000. There is an adjustment on it, if the range exceeds factory setting. Loosen the adjusting screw "A" and move the body bracket "B" toward the magneto flywheel, to increase speed. Move the body bracket away from the flywheel to decrease the speed. The bracket "C" is attached to the magneto stator. Don't make the adjustment when the engine is running.

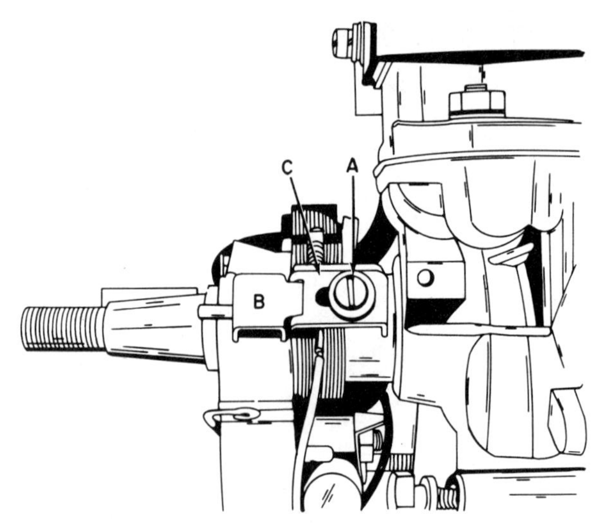

FIGURE 88. Adjusting the two-cycle Tecumseh ignition governor.

TECUMSEH FOUR-CYCLE GOVERNORS

Governors on larger Tecumseh engines are similar to the two-cycle mechanical types, differing in the details. Any defects involve some basic disassembly, but adjustments will usually control the problem. To adjust these governors on typical Tecumseh four-cycle, valve-in-head engines, follow these steps with the accompanying Figure 89: Move the remote controls to the "run" position. Loosen screw "A," move the pivot plate "B" counterclockwise and hold it. Then move the lever "C" to the left. Tighten the screw "A." The governor spring should be in the bottom center hole "D" of the plate "B." The spring should not be tampered with—stretched or cut.

Two other adjustments are possible: the maximum r.p.m. adjustment, which requires special tools (including a tachometer), and the fixed speed adjustment. Neither of these adjustments should be necessary, as they are complicated. If you believe your engine needs them, they would be best left to a service station. Such adjustments are for special equipment—for example, generators or pumps—that require specified speeds.

Governor systems in Kohler engines are mechanical types with flyweights that move by centrifugal force, either outward to increase speed or inward to reduce engine speed. Kohler says the governors should require no adjustment unless the governor arm or linkage comes loose or gets disconnected. However, if engine speed surges or changes noticeably with changing load, or if the speed drops when normal load is applied, readjustment will be needed.

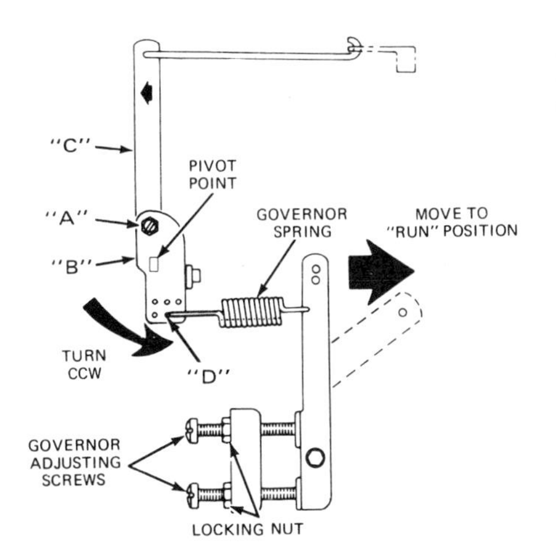

FIGURE 89.
Step-by-step adjustments on four-cycle governor.

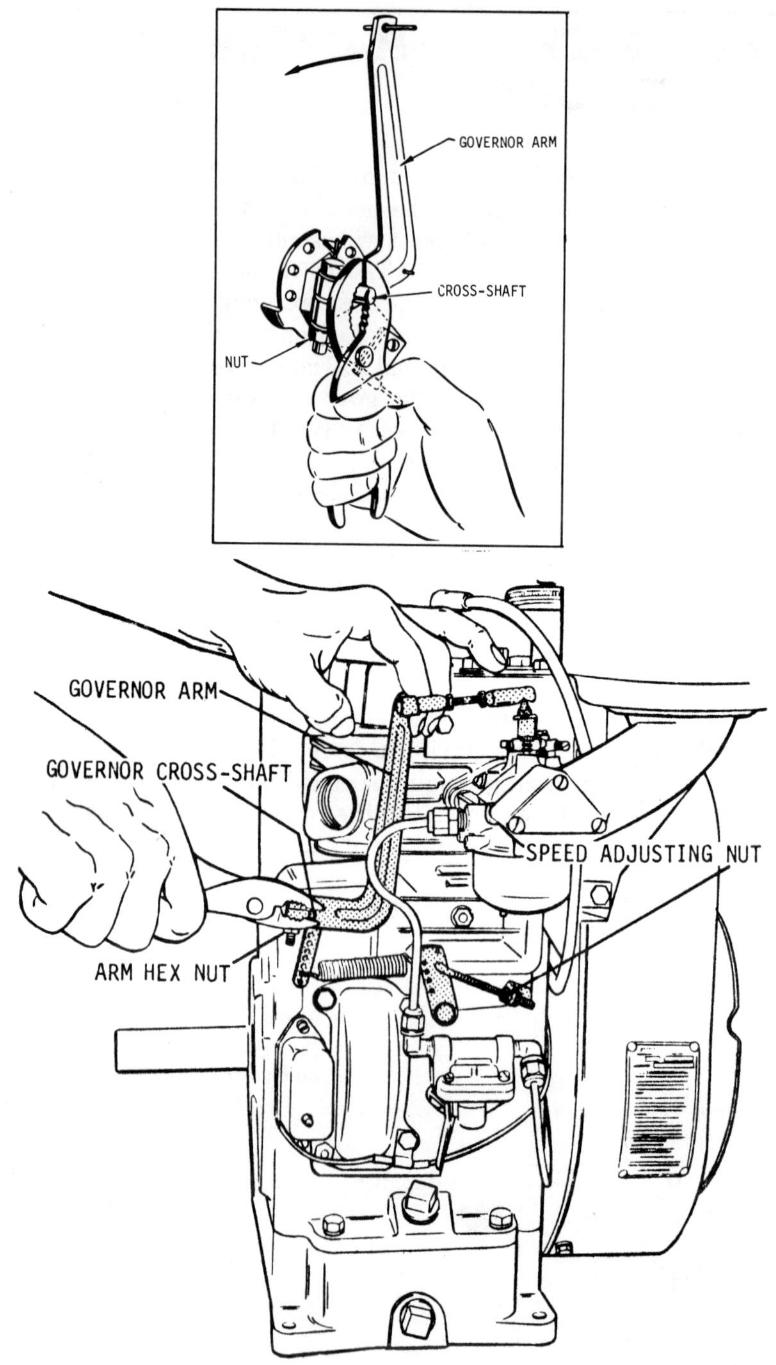

FIGURE 90. Adjustments on Kohler mechanical governors can be followed in these illustrations. Above, adjustments refer to models K91, K141, K161, and K181. Below, the initial adjustments refer to models K241 through K341.

To adjust, loosen the nut holding the governor arm to the governor cross shaft. Holding the end of the cross shaft with pliers, turn it counterclockwise as far as it will go. The tab on the cross shaft will stop against the rod on the governor gear. Pull the governor arm as far as possible away from the carburetor, and retighten the nut holding the governor arm to the shaft. (With up draft types, lift the arm as far as it will go, then retighten the arm nut.)

If the throttle wire must be replaced, it should be hooked into the speed control disk nearest the throttle bracket, with the control handle in open position. Install the cable clamp and bolt to the throttle bracket. Remove the drive pin from the speed control disk and operate the control handle, turning the disk from idle to full speed. It should move freely both clockwise to decrease speed and counterclockwise to increase speed.

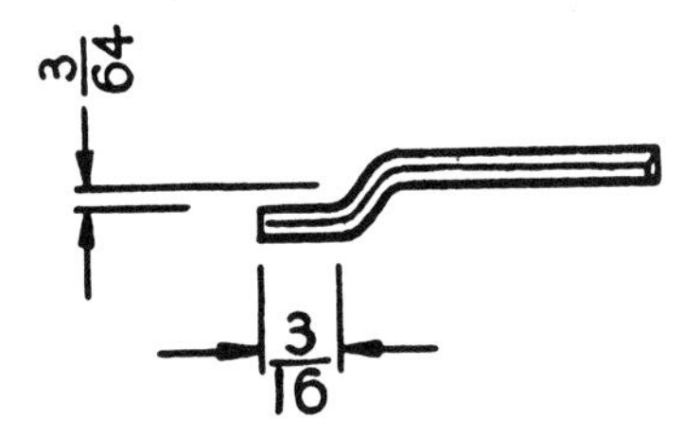

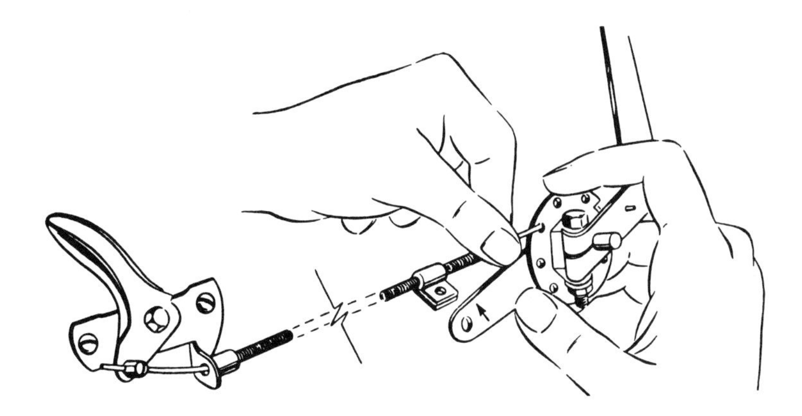

FIGURE 91. Installing the new throttle wire on Kohler models K91, K141, K161, and K181.

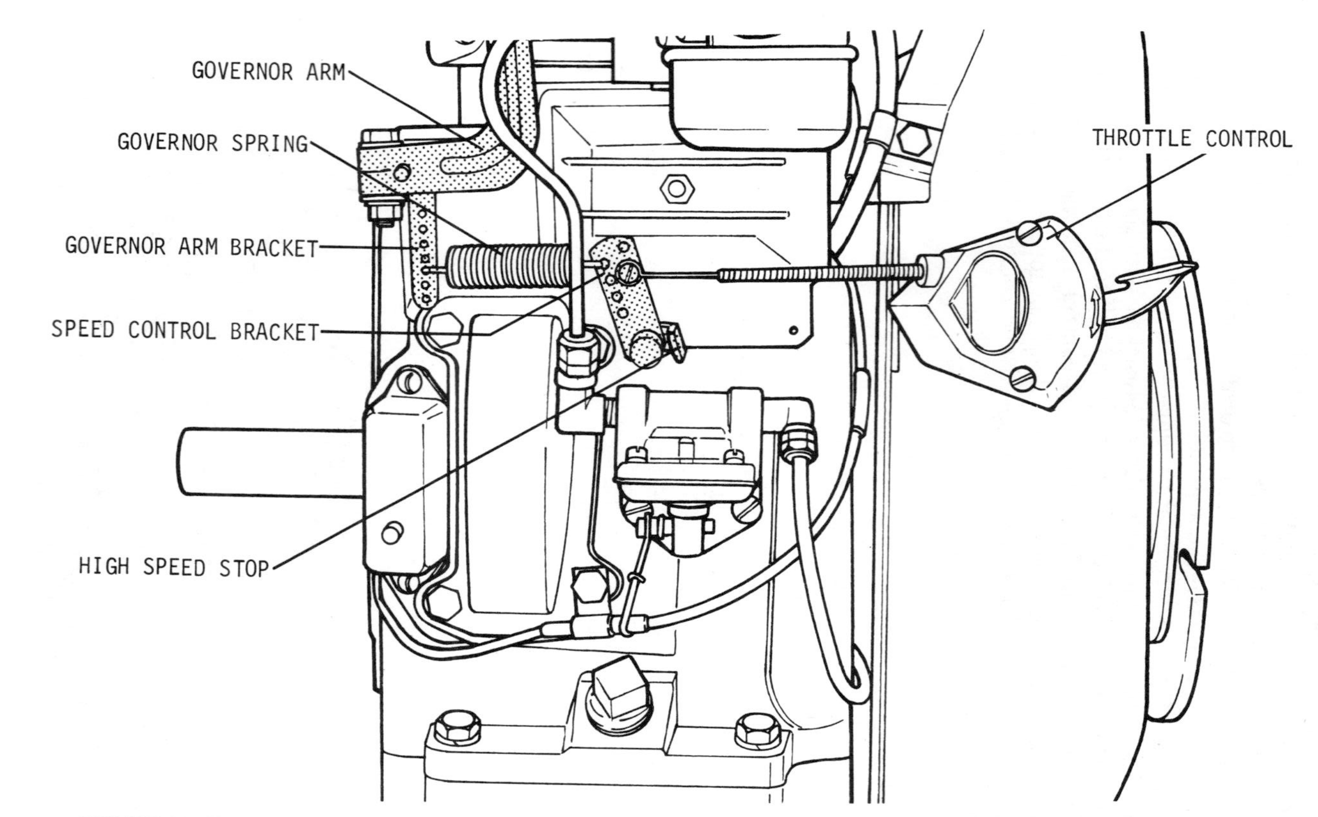

FIGURE 92. Procedures for adjusting speed on models K241 through K341 can be followed through this illustration.

Speed adjustments can be made if the engine isn't within a correct operating range—if, for example, it speeds excessively at full throttle or stalls at low speed. Loosen the large bushing nut on the governor arm slightly. Move the throttle bracket in a counterclockwise direction to increase engine speed or clockwise to decrease it. Then tighten the bushing nut to lock the throttle bracket into position. Be careful about tightening the bushing nut; excessive tightening could ruin the threads.

These directions apply to models K91, K141, K161, and K181.

To make the speed adjustment on models K241 through K341, tighten the governor adjusting screw to increase engine speed; decrease speed by loosening the adjusting screw.

Governors on Clinton and other small engines are similar enough to the devices we've discussed to make their adjustment or replacement relatively easy.

MECHANICAL STARTERS

Mechanical starters using a rope windup or a spring-loaded device are found on all small engines, but larger engines have starters similar to those on cars. These engines—which power riding mowers or other large devices—have electric starting motors and batteries or they may plug into a wall outlet for their power source.

Rope starters are the simplest and the rope is the chief wearing part. Though the rope is long-suffering, it isn't immortal. When it commences its final swooning act, the primary symptom of which is fraying, it must be replaced. Don't put it off, because it can break and give you a nasty sting in its last act of rigor mortis (the last windup).

Put the new rope through the handle (after cutting the old one off), and tie a figure eight knot. Put a pin through the knot and pull the rope tightly into the handle. Though it goes without saying, we'll say it anyway: you buy the rope according to the model number on your engine; you cannot use clothesline.

To thread the rope into the other end of the system, use a piece of music wire or spring wire, forming it precisely. Thread the wire and rope through the rope eyelet in the rope housing, out the pulley hole. The rope must pass inside a guide lug on the metal pulley. Tie a knot in the rope and pull it very tight. The knot in the pulley must not touch bumper tangs.

More recent rope starters have slightly different installation procedures. There is no guide lug, and you simply tie a knot in the rope and pull it very tight, then manipulate the knot into the knot cavity.

If the rewind assembly has become defective—unlikely in rope starters, but it can happen—and the starter housing is held in place with spot welds, drill them out with a $^3/_{16}$-inch drill. Drill deeply enough to loosen the spot welds, but no farther. Replace the spot welds with screws started from inside the blower housing, up through the starter housing mounting leg.

Older Briggs & Stratton engine rope starters may sometimes require servicing of the clutch—mostly cleaning it. Do *not* oil any part of

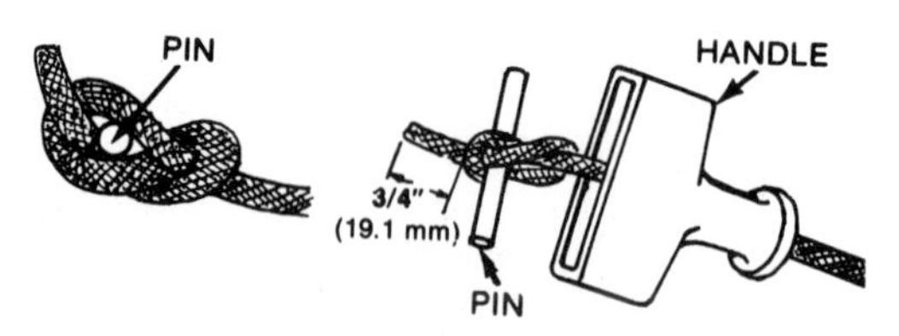

FIGURE 93. Installation of a new rope should always include sealing both ends of the knot by burning them.

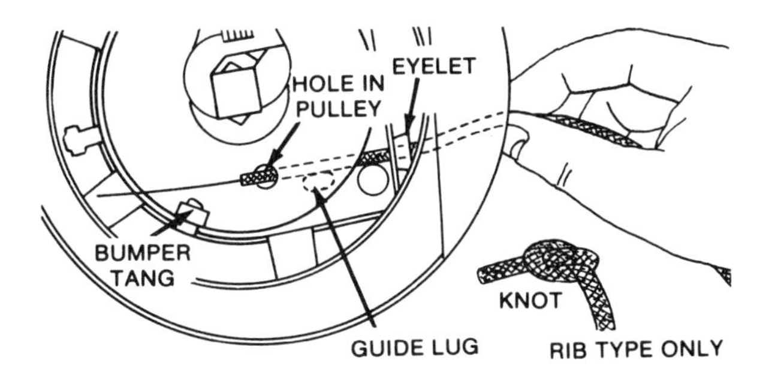

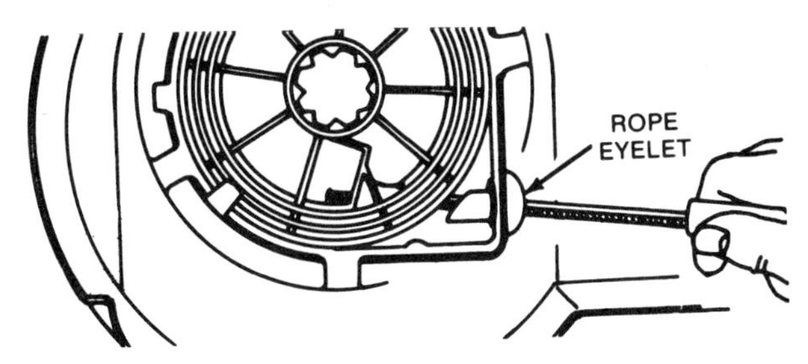

FIGURE 94. Threading wire and rope through the rope eyelet requires passing it inside a guide lug on the metal pulley.

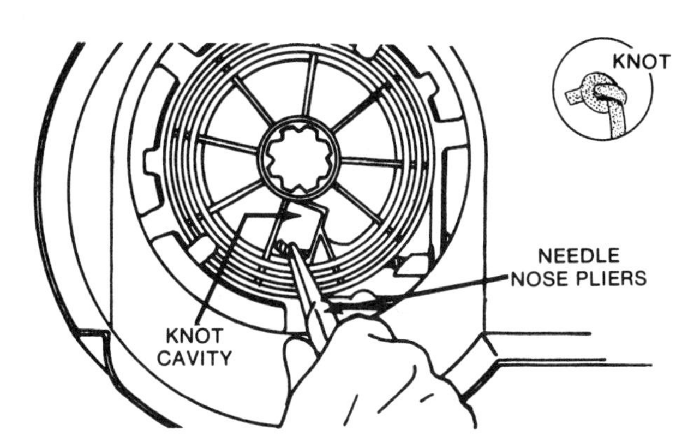

FIGURE 95. Where no guide lug exists, the rope goes like this.

it. Newer type clutches are in a sealed assembly. These should not require service, but if they do become defective, they can be taken apart. You pry the retainer cover off the housing, then clean out all parts with solvent.

If the starter spring breaks it is best to buy a new starter unit. They aren't terribly expensive, and the spring is difficult to replace.

You can tell that you have a broken spring if you wind up the starter and nothing happens; but you must distinguish between a broken spring and clutch. Watch the clutch as you manipulate the rope starter. If the clutch does not move it is probably the spring that is broken. The clutch is far less likely to be defective than is the spring; it is not a fragile component, whereas springs, even tough ones like these, wear out or break.

To remove the starter and its housing, as noted, may require drilling out spot welds.

If you want to replace the spring in a rope starter, it can be done, but don't try to replace a spring in a windup starter (one without a rope) that you crank up. If the spring breaks, buy a replacement unit. They aren't expensive and you will avoid a lot of grief and possible danger.

In rope starters, the spring can be replaced this way. Get the rope out by cutting the knot at the starter pulley. Then, with a pliers, grasp the outer end of the spring and pull it out of the housing as far as you can. Turn the spring ¼ turn and remove it from the pulley, or bend one of the tangs up and lift the starter pulley out. That will allow you to disconnect the spring without trouble.

To install a new spring, use a solvent to clean out the rewind housing as well as the pulley and the new spring, and wipe all the cleaned parts thoroughly. Oil the spring and insert either end of it into the blower housing slot. Hook the spring into the pulley. Put a bit of grease on the pulley. The tang you lifted up must be bent down, adjusted to a 1/16-inch gap—and bent down to that gap. The pulley must be pushed entirely into the rewind housing when you make that adjustment and measurement. The big problem now arises—getting the spring wound up. To do that—and we approach the reason it is best to buy an entire new component that includes the wound-up spring—you must use a ¾-inch square piece of wood. You place that block into the center of the pulley hub. (It will fit there.) Then, with a wrench, turn the wood block counterclockwise until the spring is wound tightly. Then back off the pulley one turn, or until the hole in the pulley that is for the

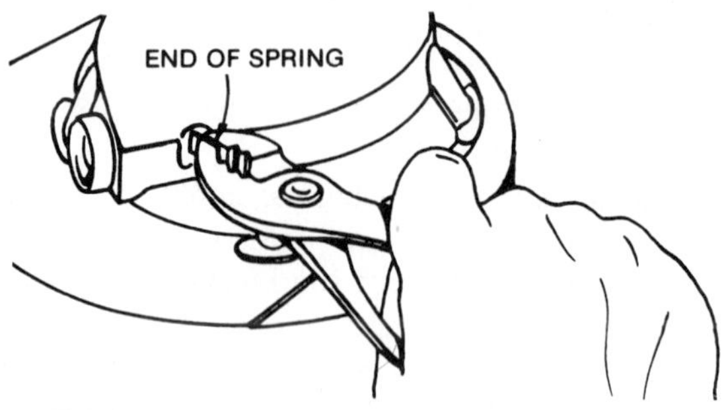

FIGURE 96. Removing spring from a rope starter.

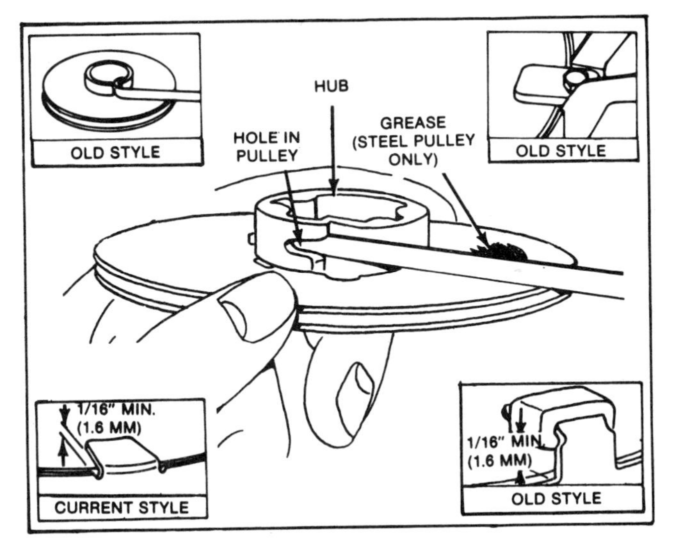

FIGURE 97. To install spring and adjust tang, insert either end into the blower housing slot and hook it into the pulley.

rope knot aligns with the eyelet in the blower housing. Lock the end of the spring into the smaller portion of the tapered hole. That will do it.

Thus far we've been discussing Briggs & Stratton engines with the starter on top. But the vertically arranged starter is also common, and different. It isn't very different, but just enough to require different instructions.

To remove and install a new rope on a vertical rope starter on Briggs & Stratton engines, use a screwdriver to lift the rope out about a foot. Wind the rope and pulley counterclockwise two or three turns, and that will release tension on the spring. Then pry off the cover, using a screwdriver. Next, remove the anchor bolt, anchor and rope guide. Mark the position of the link and then remove the assembly from its housing. To insert the new rope, make a tool or simply use a needle-nose pliers. Clean out all the parts and check out the operation of the link. It should move the gear to both extremes of its travel. If it does not, you need a new one.

Installation of a new spring in these vertical pull starters is much easier than in the other types. Hook one end in the pulley retainer slot and wind it until the spring is coiled into its housing.

Thread the rope through its grip and into the rope insert. Tie a small knot as tightly as possible. Briggs & Stratton recommends that you apply heat to the knot to seal it, but it doesn't have any place to go once you get it in place. Pull the knot into its pocket and snap the insert into the grip. Push the rope through the housing and into its pulley, and here you'll need some kind of rope inserter tool made from a wire. Put the pulley assembly into its housing with the link in its pocket, and

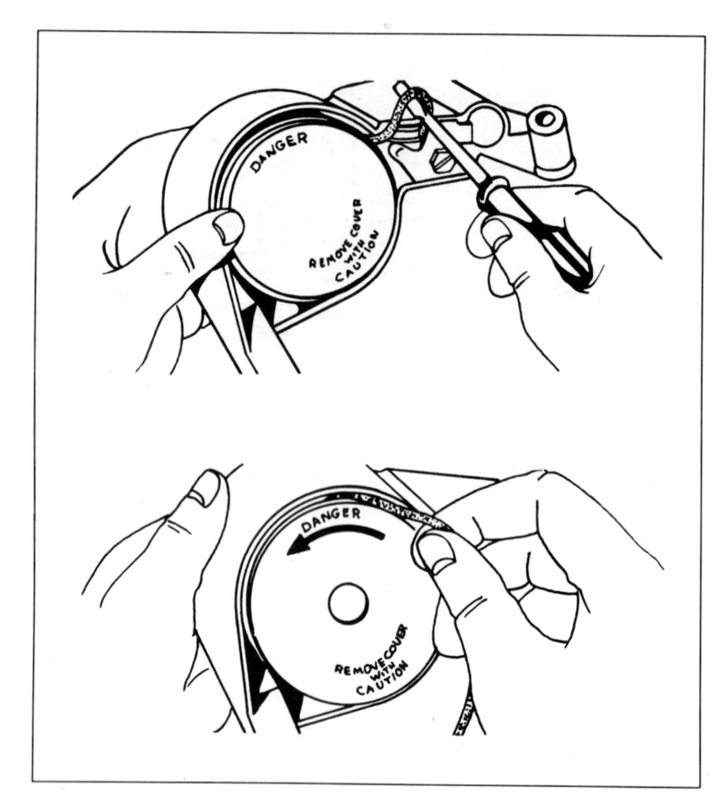

FIGURE 98. Pulling up the rope.

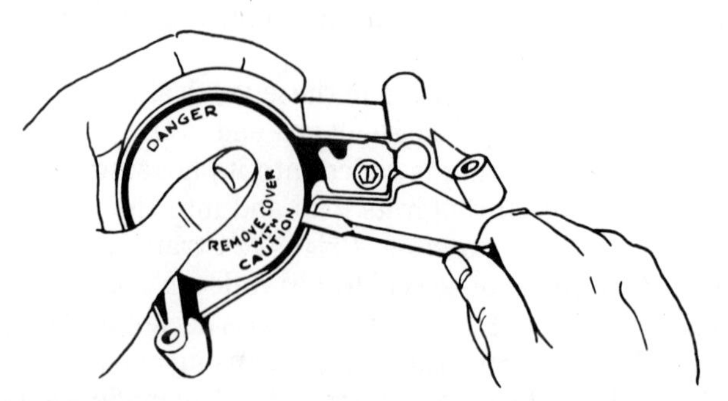

FIGURE 99. Prying off the cover of a Briggs & Stratton vertical pull starter.

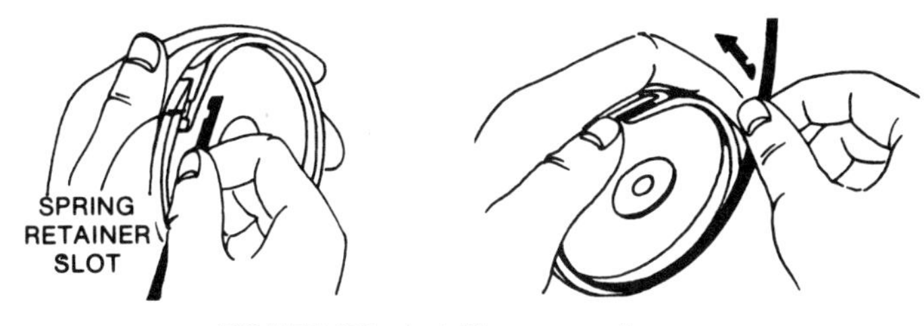

FIGURE 100. Installing new spring.

install the rope guide. Rotate the pulley counterclockwise until the rope is where it belongs. Now hook the free end of the spring to the spring anchor and install the screw—tightly. Put some engine oil on the spring. Snap the spring cover into place. Wind the spring by pulling the rope out about one foot, then wind the rope and pulley two or three turns, clockwise, to get the proper rope tension.

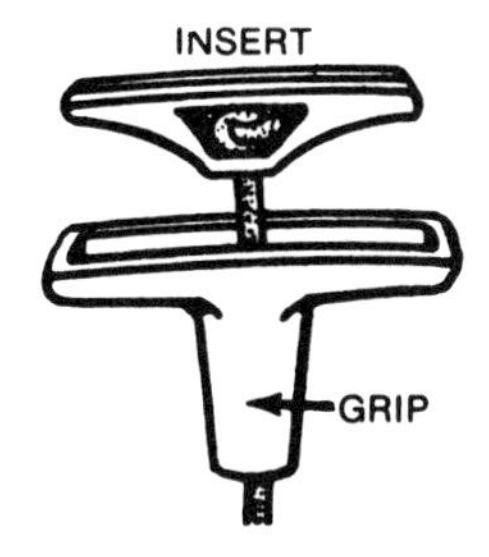

FIGURE 101. Threading the rope into its grip.

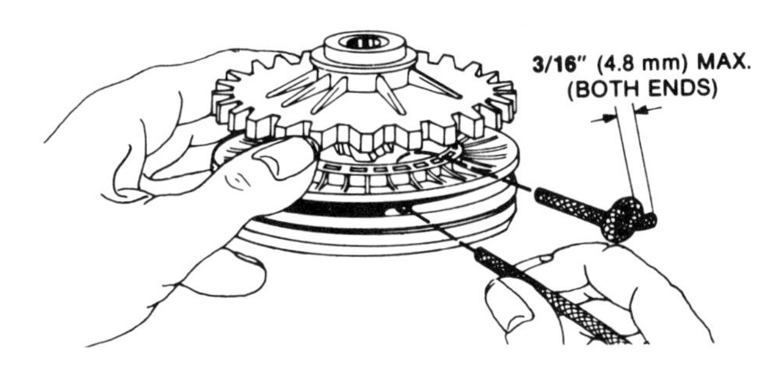

FIGURE 102. Putting the rope in its pulley.

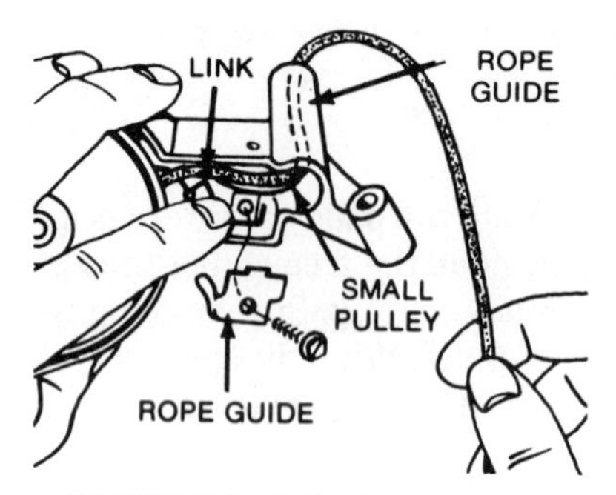

FIGURE 103. Pulley installation.

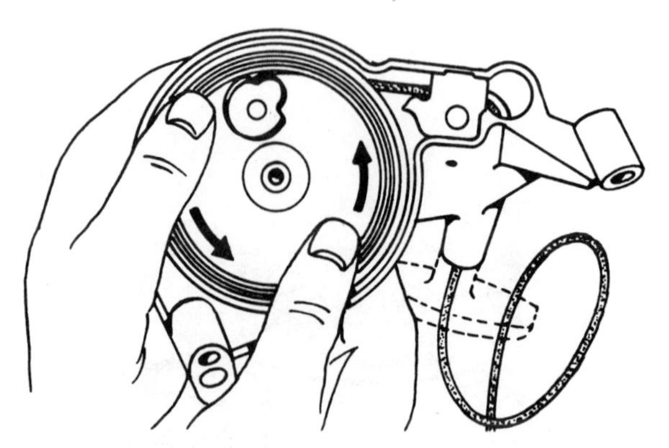

FIGURE 104. Getting rope tension regulated.

TECUMSEH REWIND STARTERS

Tecumseh rewind starters have their own tricks and complexities though they have many of the same characteristics as those units we've been discussing. To take a rope starter apart you untie the knot in the rope, while slowly releasing the spring tension. If you are replacing a broken rope, the tension will have been released, and one usually replaces a rope that shows signs of needing replacement. Since the spring will be under tension, you will need some kind of holding device on the rope—say a vise grip pliers—below the knot as you untie the knot or cut it off. Remove the retainer screw, retainer cup, the starter dog and spring, and the brake spring (and the cam dog on the Tecumseh snowproof engines; that is, the retainer or clamp, which, in turn, has another retainer). Lift out the pulley, then turn the spring and keeper assembly to remove it.

Once you replace any worn parts, you assemble the starter as follows. Place the spring and keeper assembly into the pulley. Turn it to get it locked in position. The spring should have grease or oil on it. Next, the pulley goes into the starter housing. Then install the spring, the starter dog, and the dog return spring. Replace the retainer cup (and

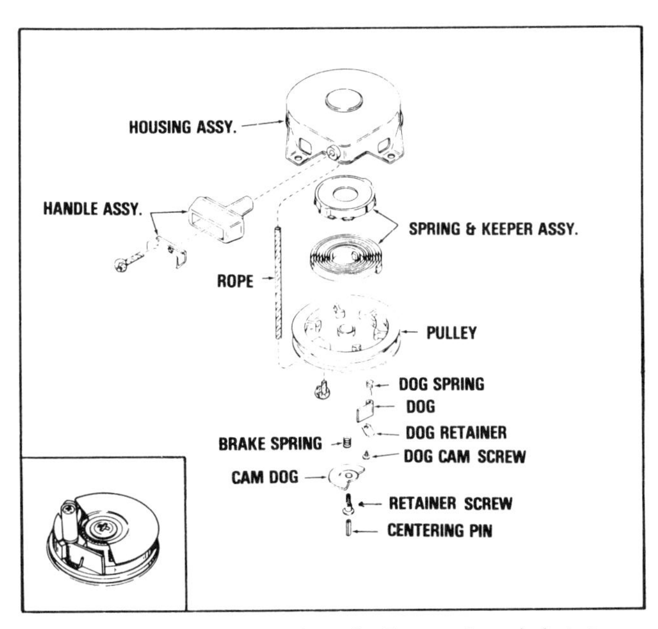

FIGURE 105. Exploded view of a Tecumseh rewind starter.

the cam dog on the snowproof type) and the retainer screw. Wind the pulley counterclockwise until it is tight, then allow it to unwind until the hole in the pulley lines up with the eyelet in the housing. Install the rope and its handle. Use a No. 4½ or 5 braided rope. Burn the ends with a match and wipe them off with a cloth while hot. Rope length should be 54 inches. Check the length of the old rope. Next, the gear and pulley go together, followed by the washer and snap ring. Put grease on the center shaft, place gear and pulley in position, with brake spring loop positioned over the metal tab on the bracket. The rope clip fits tightly on the bracket; the raised spot fits in the hole in the bracket. Hub and hub screw are next to be installed, then install the spring. It comes in its own retainer. Lay the spring and its retainer over the receptacle and push it out of its retainer into the correct position. Install the cover.

Wind the rope onto the pulley by slipping it past the rope clip. When it is on the pulley, wind it two more turns, thus putting tension on the spring.

In mounting the starter assembly, place it so that the head of the tooth is not closer than 1/16 inch from the base of the flywheel gear tooth.

Tecumseh vertical pull starters differ from the above in various details. To disassemble, pull the rope out sufficiently to lock it into the "V" of the bracket. Put the starter bracket over a socket wrench large enough to receive the head of the center pin and drive it out, taking care not to mangle it. Rotate the spring capsule strut until it aligns with the legs of the brake spring. Put a nail through the hole in the strut so it catches in the gear teeth, thus holding the capsule in a wound position. Then slip the sheave out of the bracket. Don't remove the spring capsule from the sheave assembly unless it is fully unwound. To do that, squeeze and hold the spring capsule firmly against the gear sheave with your thumb at the outer edge of the capsule. Then slowly take out the retainer nail from the strut and relieve the grip on the assembly. That allows the spring capsule to rotate under control and to be completely unwound.

Reassembly is the reverse. Vertical Tecumseh starters use No. 4½ braided rope. The length with handle mounted on the blower housing is 65 inches. Put the rope end through the hole of the gear sheave opposite the staple platform and tie a left-hand knot. Pull the knot back into the rope handle cavity. The rope end should not protrude. Wind the rope onto the sheave clockwise, as viewed from the gear side of the assembly. Install the brake spring and capsule, hooking the spring on the gear hub. Wind the spring four turns and align the brake spring legs with its strut. Insert the nail back into the strut. Some of these starters have a locking device (a "pawl") or a delayed locking device and spring. They go on next—if you've taken them off, now you must contend with their assembly. Feed the rope end under the rope guide and hook it into the "V" notch and remove the nail. The strut will rotate clockwise against the bracket. Press in or drive in a new center pin.

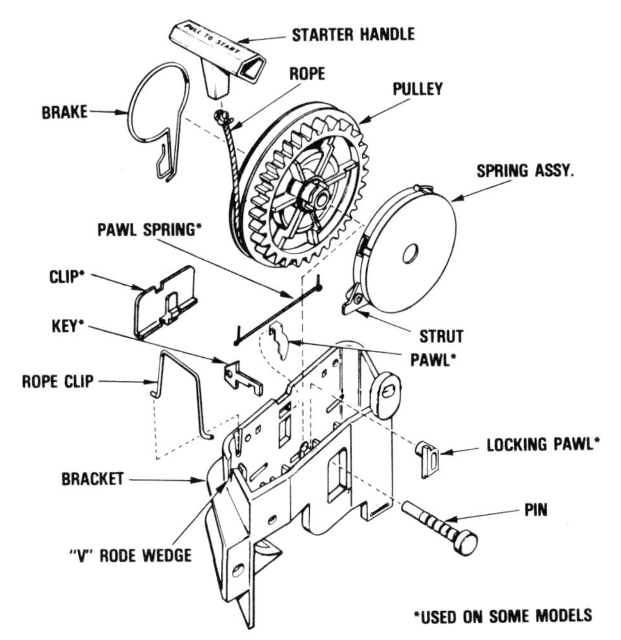

FIGURE 106. The Tecumseh vertical pull starter has its own details of disassembly and assembly.

Tecumseh starters with "V" notches in the bracket allow you to change the rope without taking the starter apart. Just turn the pulley until the staple in the pulley lines up with the notch. Pry out the staple with a screwdriver and remove the old rope. Fully wind up the spring, then let it unwind until the hole in the pulley—180 degress from the original staple mount—lines up with the notch. Feed the rope through the hole and tie a left-handed knot. The rope end must not protrude from the knot cavity.

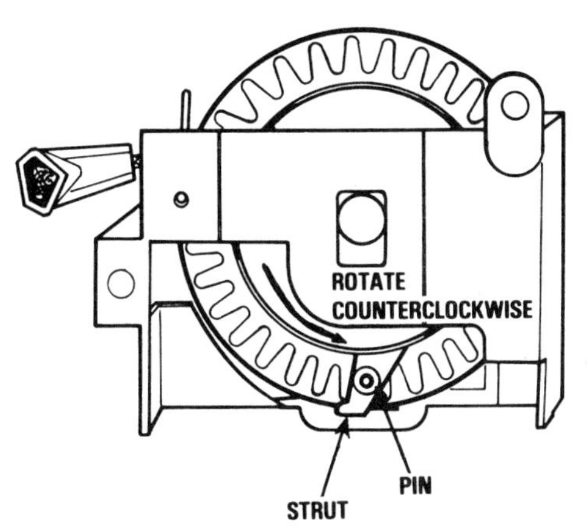

FIGURE 107. Tecumseh vertical pull starter rope adjustment.

KOHLER STARTERS

Kohler starters, using rope or rewind springs, operate about like all the others discussed. That is, they require elbow grease rather than electric motors to turn the engine over. Nonetheless, Kohler rope or spring windup starters differ in many details, including some interesting caveats about their use. Anyone accustomed to dealing brusquely with rope starters may be put off by the rather effete requests made by Kohler. Thus, after the engine catches, "do not allow starter rope to snap back into starter housing. Continue to hold handle and allow starter rope to rewind slowly." Failure to observe this request can result in a shorter life for the starter. Kohler has no insight into what this advice will do for you. It goes on to remonstrate on the delicacy of the matter: "Do not use starter in a rough manner, such as jerking or pulling starter rope all the way out. A smooth steady pull will start engine under normal conditions." If that isn't enough, Kohler specifies the way you use the starter handle: It is to be pulled "straight out so that rope will not receive excessive wear from friction against guide."

If the starter rope breaks or the spring fails, the repair goes as follows. Four blower housing flange bolts must be removed before the starter can be taken off. There is a trick to this; a retainer ring must come off using a screwdriver, but you must also use your thumb to hold the washer in back of it; otherwise, a spring will push the washer out once the retainer is off, and such parts always disappear into the void.

Remove the washer, spring, and other washers, along with the so-called friction shoe assembly, which consists of some springs and fiendish little gizmos whose function it is to regulate braking and releasing. On model 425 only, relieve rewind spring tension by removing the handle from its catch and then allowing the rotor to unwind slowly. On model 475 only, hold the rotor and gradually release it. Prevent the rewind spring from getting out of its cover by carefully lifting the rotor about ½ inch and detaching the inside spring loop from the rotor. If you let the spring get out you can replace it in the cover simply by coiling it up.

Once everything is out—presumably because the rope or spring needs replacement, or there are worn parts that prevent free starter operation—cleaning should be done thoroughly. Then you reassemble, with whatever new parts are required. Don't replace parts unless they are broken, bent, worn, cracked, or clearly don't work. Springs and ropes are the usual defective items, though anything designed and made by man requires replacement sooner or later, including the designer-maker.

Kohler identifies two types of these hand or "retraceable" starters, the Fairbanks-Morse and the Eaton. They are similar. The Fair-

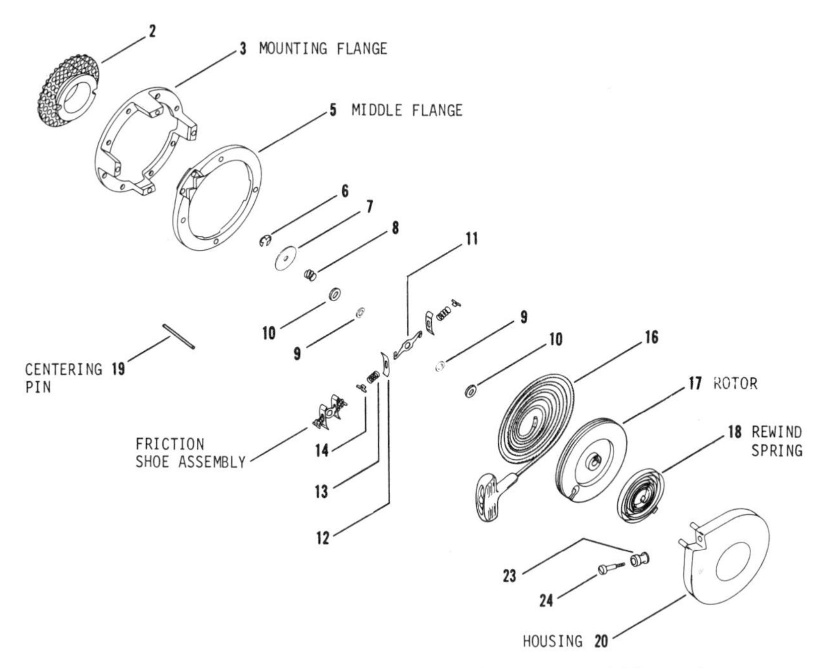

FIGURE 108. Fairbanks-Morse retractable starter on Kohler engines.

banks-Morse rope replacement procedure requires that it be threaded through the rotor hole provided for it; you then wind the rope on the rotor. Replace handle and washer (if there is one), and tie a double knot in the end of the rope at the handle.

To rewind the spring, start with the inside loop and carefully remove the old spring from the cover by pulling out one loop at a time. Replacement springs have a holder to simplify the assembly. Put the spring in position (as shown in illustration D in Figure 109), with the outside loop engaged around the pin. Press the spring into the cover area, thus releasing the spring holder. Put a few drops of oil on the spring and some light grease on the cover shaft.

Place the rotor, complete with cord and handle, into the cover (#20 in Figure 108) of model 475 only, and hook the inside loop of the spring to the rotor, using a screwdriver (Illustration E of Figure 109). On model 425 the rope must be completely wound up in correct rotation on the rotor before installing it in the cover. Next, replace washers #9 and #10 (in Figure 108), the friction shoe assembly, and the second set of washers #9 and #10, spring #8, washer #7, and the retaining ring #6. Model 475 ("pre-tension") requires four additional turns of the starter rope. Also, model 425 needs the rotor turned five additional turns with the aid of the cord—in the same direction as the rope goes.

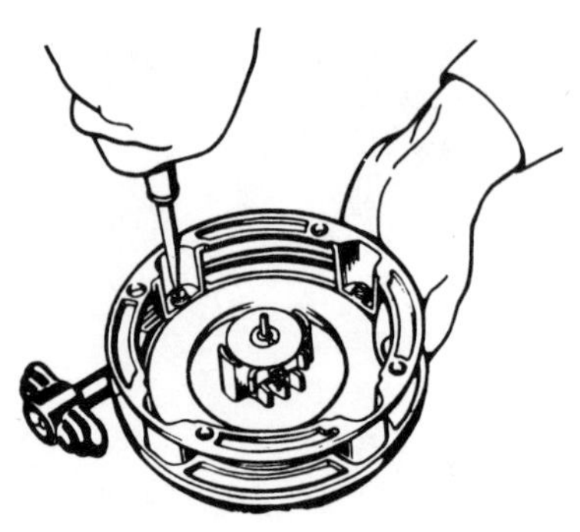

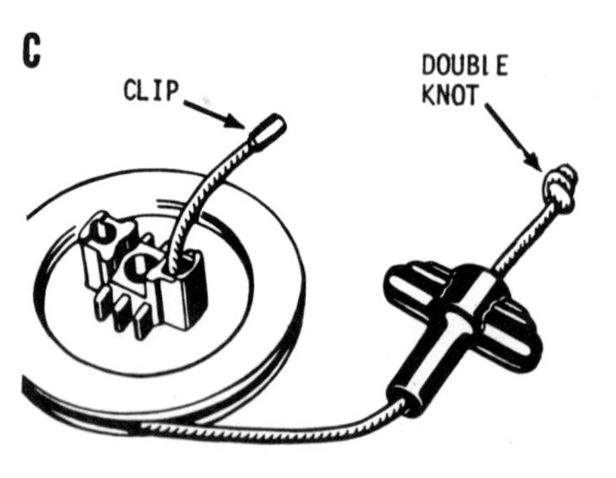

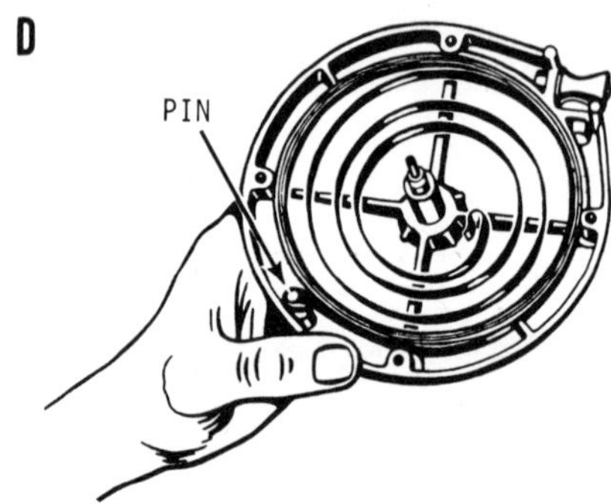

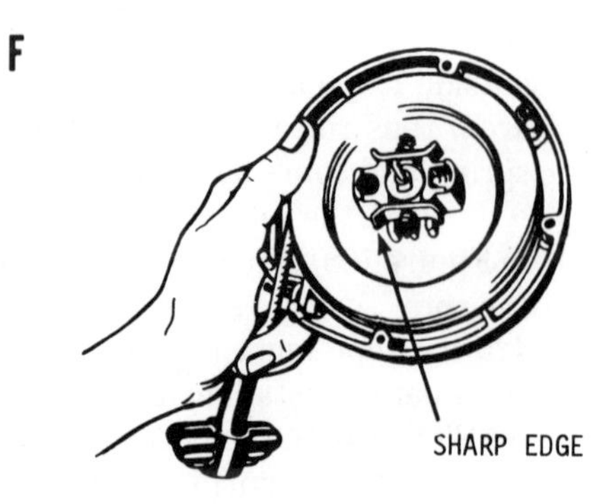

FIGURE 109. Fairbanks-Morse repairing on Kohler engines.

In 475 and 425 models, hold the rotor as shown (Illustration B in Figure 109) with a screwdriver, and replace flanges #5 and #3 (Figure 108), also screw #4.

The starter must be centered. Pull out the centering pin #19 about ⅛ inch. Place the starter on its four screws, getting the centering pin to engage the center hole in the crankshaft. Press the unit into position. Hold the starter with one hand, put the lock washers and nuts on the screws with the other hand, and tighten them securely.

Align the starter on the engine, with the centering pin engaged in the center hole of the crankshaft. If the pin is too short to reach the crankshaft, pull it out with a pliers. Press the starter into position and install the four screws with their washers—both lock and flat washers. Tighten securely.

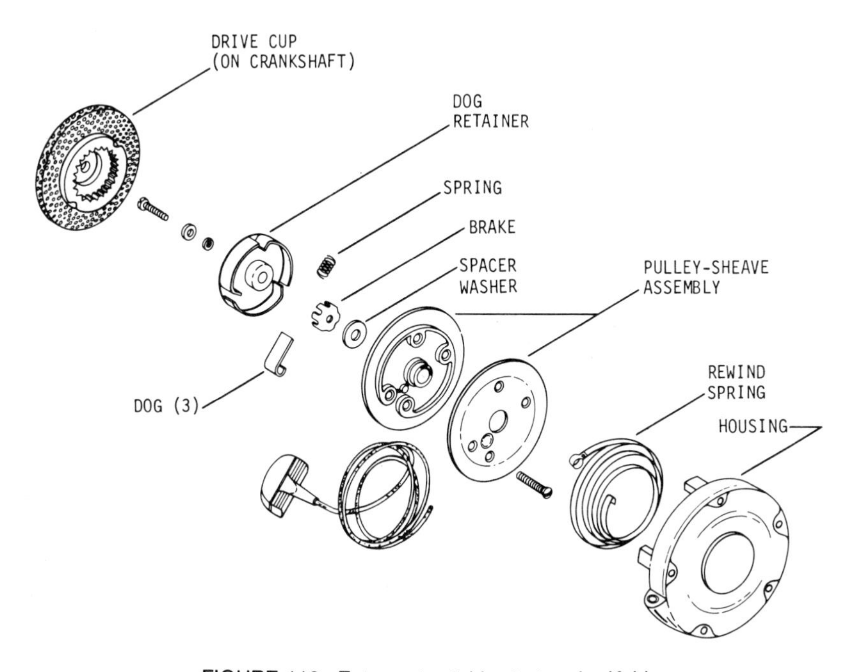

FIGURE 110. Eaton retractable starters by Kohler.

EATON STARTERS

The Eaton starters require service only if the rope or the rewind spring fails. Five mounting screws hold the assembly to the blower housing. Be careful, when disassembling, that you don't disturb the spring until you are ready for it.

Remove screw and washers on the dog retainer. Slip retainer off the small spring that is fastened over the post on the outside face of the pulley. When you take the retainer off, do it very gently so as to avoid damaging the spring.

Now, to relieve the main spring tension, pull the rope handle out about eight inches and tie a knot in it to prevent the rope from being pulled into the housing. Put a screwdriver blade under the rope retainer on the handle, slip the rope out of the retainer, and untie the knot at the handle. Hold the pulley-sheave assembly with your thumb to prevent the rewind spring from cutting loose rapidly; untie the other knot and slowly let out the spring.

You aren't home free yet. Slip the pulley assembly out of the housing, being aware that the inside loop of the spring, which fits into the inner hub of the assembly, can unwind violently unless it is held in the housing during its removal. No special trick is involved other than holding it in place.

To replace the starting rope, remove the four screws on the sheave side of the pulley assembly. Separate them, take off the old rope, put on the new, and tie a double knot in the end. Put the pulley and sheave back together and rewind the rope. Nothing to it. Notice how the old one comes off so that you can put the new one back exactly the same way.

In the unlikely event that the rewind spring must be replaced and you resolve to do it, remove the spring, one loop at a time, starting with the inside loop. Pull the spring out, grasping it with a vise pliers or something similar. Then it is easily controlled. Put the new spring into the housing, block it so that it can't jump out laterally, and remove the retaining clip and tape. If there is tape, cut it off in segments; don't peel it and risk disturbing the spring.

To reassemble the housing, bend a piece of wire to form a hook. Then catch the inside loops of the rewind spring and pull it out sufficiently to allow the hub on the inside of the pulley to slip into the inside of the spring. Slide the pulley-sheave into its correct position with the hub inside spring. Remove the wire, fully seat and turn the pulley until the spring engages in its slot on the hub. Replace the dogs in the pulley, replace the spacer washer, and hook the spring over the pulley shaft. Install retainer and washers, and tighten the screw.

Now the spring must be tensioned. Put the end of the rope through its bushing in the housing and pull the rope out until the notch in the pulley is aligned with the bushing. Hold the pulley and let up on the rope; then place the rope in the notch, and after blocking the housing to prevent it from turning, rotate the pulley and pull the slack rope through the bushing, thus adding tension to the spring. Tie a knot in the rope, temporarily, to hold tension while installing the handle.

Thread the end of the rope through the handle, then through the rope retainer. Tie a permanent knot in the end of the rope and reinstall the retainer in the handle. Untie the temporary knot, pull the rope to its fully extended position and release it. The rope should fully rewind until the handle hits its housing.

No service should be required of these starters, other than rope replacement, and, less often, replacing the rewind spring. But it is important to replace the rope when it begins to fray; don't wait until it breaks completely. If it breaks, the pulley will unwind violently, possibly breaking the spring and other components as well.

If the rope is only frayed, pull it to its fully extended position and secure the pulley in this position, blocking it to prevent rewinding. Cut the knot off and remove the rope. Put the handle on a new rope and slip the other end in through the bushing in the housing and the hole in the pulley. Put back the rope retainer washer, then tie a knot in the rope. Burn the end slightly to fuse it. Slowly release the pulley, braking it so that the rope winds slowly around the pulley until fully retracted.

Now it is necessary to align the starter to the engine. Attach it with retaining screws, but don't tighten them much. Pull out the starter handle about eight inches so that the starter centers as the dog retainers engage in the drive cup (you can feel it). Hold the rope in this position and tighten the retaining cap screws.

ELECTRIC STARTERS

As you get into heavy equipment—riding mowers, snowblowers with big capacity, etc.—you will find electric starters. Two types will be encountered, 12- and 120-volt systems. The 12-volt systems are more or less identical with automobile motors and attendant systems; the 120-volt systems operate from a house or garage electrical outlet. The latter type means that you don't want an engine that stalls too far from the outlet, unless you enjoy strenuous pushing. When tuning the engine in such equipment be especially conscientious. The 12-volt systems may have batteries on board, which means they also need a generator to keep the battery charged. In fact, the variety of electric starting systems is perplexing—unlike car systems, which are all alike. One type of electrical starting system, for example, combines the starter and generator into a single motor; others have separate chargers that are independent chargers having no special relationship with the starting system. There are also two kinds of drive systems: (1) belt and pulley such as you find in

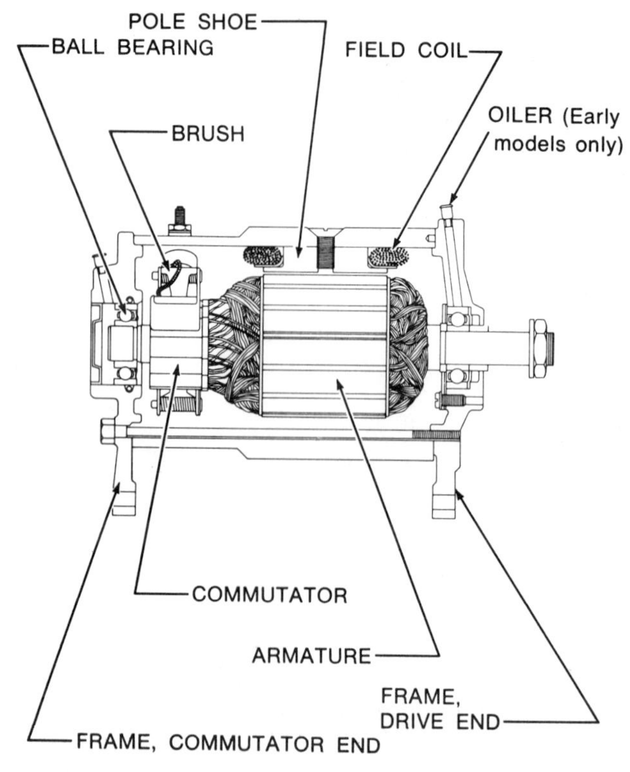

FIGURE 111. Kohler motor-generator cutaway view.

power tools, furnace blowers, etc., and (2) pinion gear drives like those in automobile starters.

To keep all this in perspective, we deal with motors, their connections, wiring, connecting gears, and power sources. There is nothing mysterious, but every component is subject to wear.

Kohler makes a starter system that combines the motor with the generator. It uses a 12-volt battery, a voltage regulator, the motor-generator, and a belt drive system.

The motor-generator has both cranking and generating windings. The cranking winding is a heavy-gauge, low-resistance wire designed to carry as high a current as possible. During cranking the current from the battery flows through this circuit, creating a high-density magnetic field that interacts with the armature windings and forces the armature to rotate. After the engine starts and the starting switch opens to discontinue the cranking circuit, the generating or shunt winding takes over and functions as a conventional generator, as the shunt field produces energy for recharging the battery.

These motor generators may be mounted at the flywheel end of the

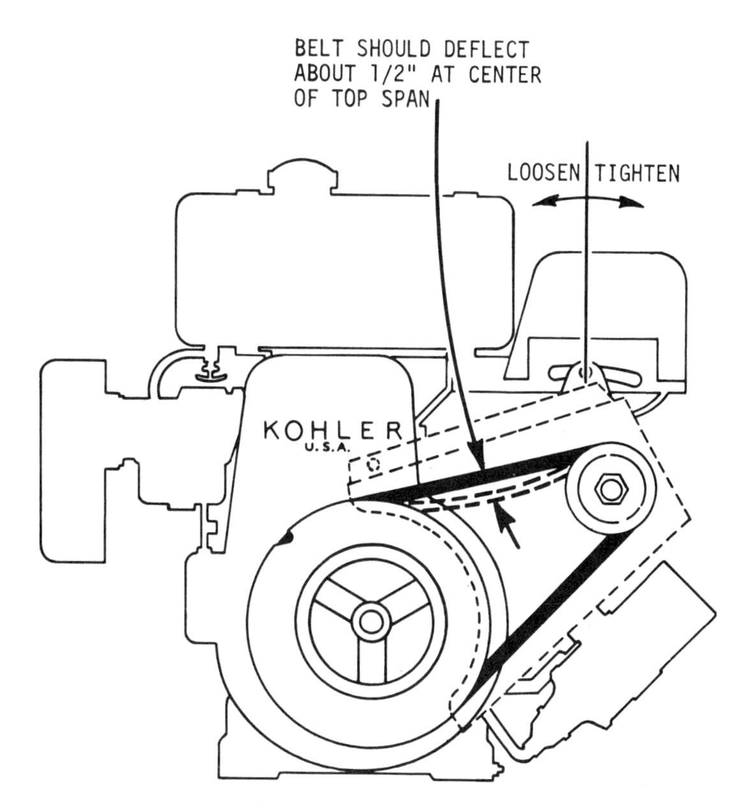

FIGURE 112. Belt drive on Kohler motor-generator.

engine and rotated in clockwise direction, or they may be mounted at the rear and rotated in the reverse or counterclockwise direction, as viewed from the pulley end. Kohler warns against using the starter for more than 30 seconds at a time without pausing to cool it for at least two minutes. Overheating can ruin the unit.

The drive belt should be checked at least once a season for tension and wear. If you can push it in more than a half-inch it needs tightening. Examine its inner surface for cracks and fraying. To adjust, loosen the cap screw that holds the motor-generator to the upper bracket, then move the unit until the desired tension is obtained. Tighten the adjusting screw.

Kohler recommends periodic inspection of the motor if it is operated in dusty or dirty conditions, at high temperatures, or at continuously full output. Moisture can also cause trouble. Inspect the mounting and wiring connections for tightness and cleanliness, then for internal motor problems.

Sooner or later, if the motor gets much wear it will burn out brushes and commutator. These are both constantly wearing components of every electric motor or generator. You can inspect brushes by removing the through-bolts and the commutator end of the motor frame. You inspect for brushes that are worn more than half-down, but you can't tell that unless you see a new brush. So buy some new brushes. When you inspect, don't pull the end of the frame all the way out, unless you are planning to put new brushes in. Getting the frame back on can be a pain, and you don't want that pain unless you also get the benefit of the new brushes. So look. Also examine the commutator; wear on it will show up as roughness and unevenness, dirt and mica insulators that are flush with the copper commutator segments. In that case, the mica must be undercut, using two files—one three-cornered, the other straight edge.

When replacing brushes, replace the brush springs. Usually they are sold as a unit. Even if the older spring appears to be O.K., it almost certainly isn't. Springs lose tension. That causes poor contact between brush and commutator, thus causing arcing (short-circuiting) and rapid wear of both brush and commutator.

Replacing brushes can be ticklish in any electric motor. You have to improvise a way of keeping the brushes drawn in their holders sufficiently to allow the commutator to slide between them back into its housing (bearing). Use string, wire, or anything that strikes your fancy. Sometimes you can bend a paper clip and push it between brush and holder, to prevent the brush from being pushed out by spring tension. These are all delicate parts and you must not use force on them.

Sometimes the coils burn out. Usually that means the motor won't run at all. Sometimes an armature burnout is only partial, affecting one series of coils. The motor will run, but badly. In either case, new coils

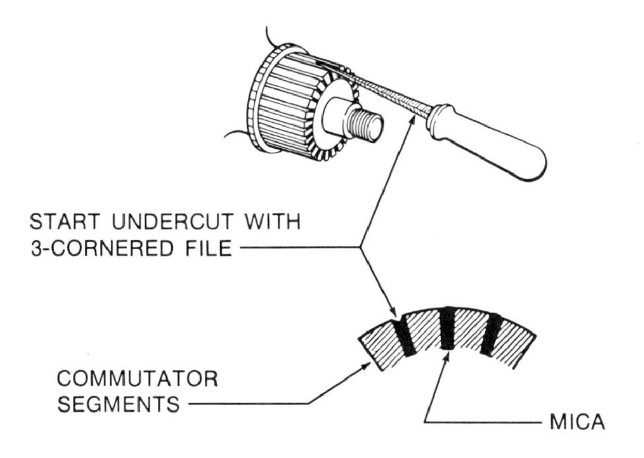

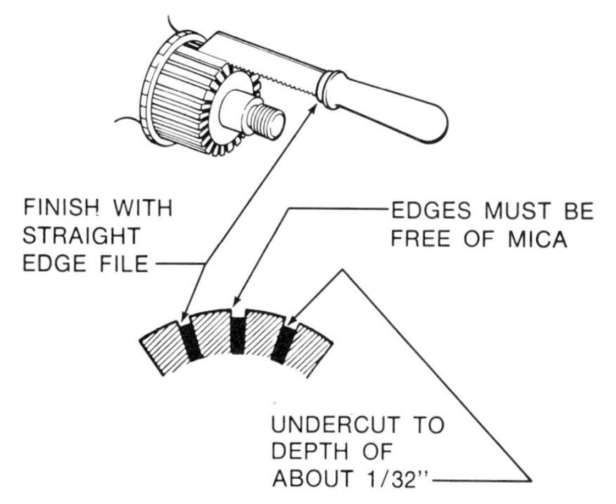

FIGURE 113. Undercutting mica on motor-generator (or any other) commutator. You can renew any electric motor like this if the coils check out.

and components are indicated. You have to submit the problem to a motor rebuilding shop, and an armature that has defects in the coils usually will not be repaired as it is too costly. Shops test armatures (rotors) with something called a growler, which picks up short circuits. It's not a piece of equipment you're apt to have under the kitchen sink, nor is it one that you should rush out to buy.

Whenever a motor is taken apart, bushings and bearings should be cleaned out and greased—or replaced if you see signs of excessive wear, such as cracks, flat spots, rough spots, or looseness of fit. If bearings are of the ball type, they should be soaked in a cleaning solvent. Sleeve bearings can be wiped out. Apply heavy-duty grease to such bearings, but don't overdo it; the grease can be forced out and cause trouble.

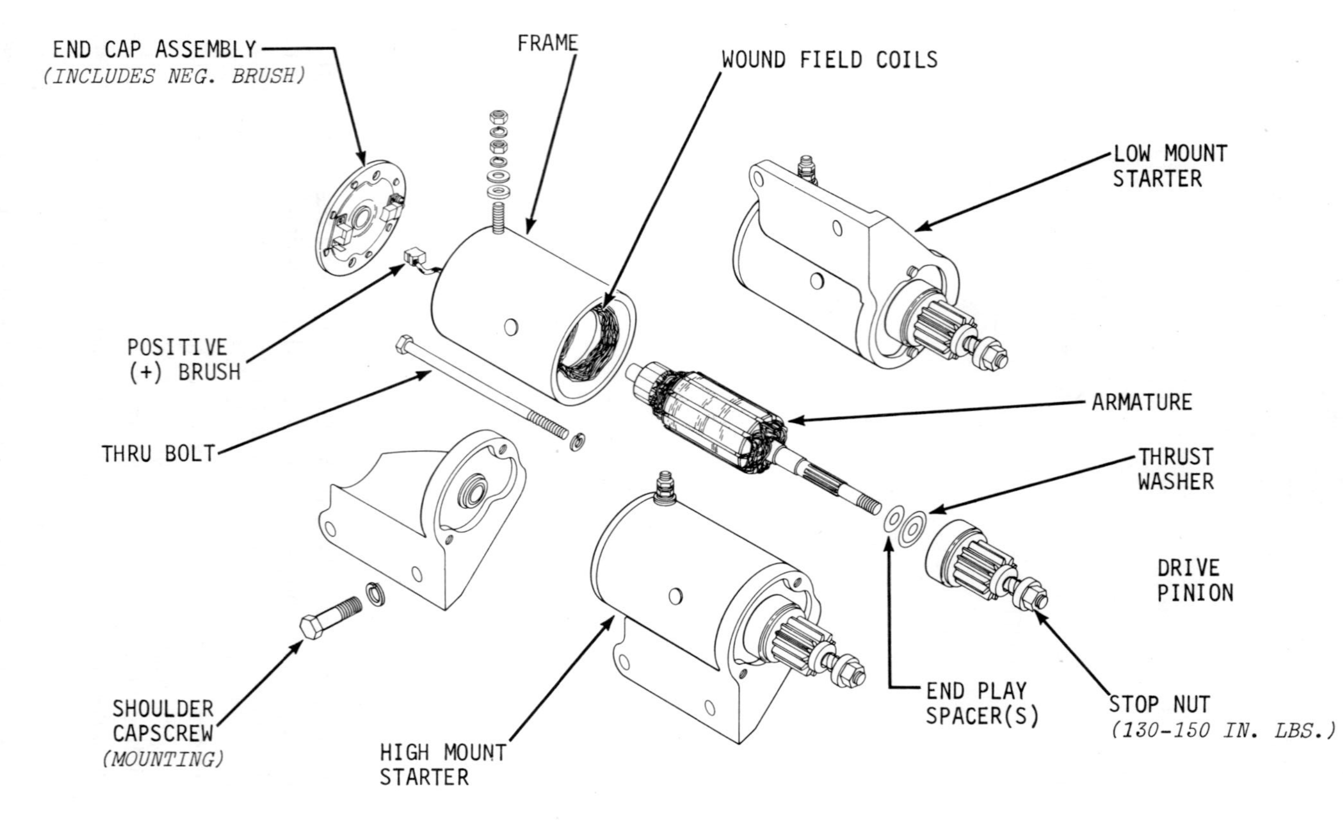

FIGURE 114. Wound field coil motor used in Kohler starters.

Some motors have cap oilers for the bearings. These caps pull up and you add specified oil—usually engine oil—to the receptacle. Usually only a few drops, periodically, are prescribed. In this case you don't grease the bearings.

The most common system in use on heavy-duty motors is either grease and/or oil, or "permanent" lubrication. The definition of permanence is anybody's guess. Motors under heavy use require lubrication at some interval. No matter what name is attached, ("permanent," etc.), no grease suddenly becomes permanent. So, anytime you take a motor apart for whatever reason, and it has "permanent" lubrication, it is well to add a little impermanent grease to the bearings, if you can get at them at all. It sometimes takes a bit of doing. Permanence in these matters evidently relates to the permanent prevention of grease getting in where it is needed.

The combination of starting and generating within one unit, as described earlier, places a double strain on all systems of the device. These strains are both mechanical and electrical.

Disassembly procedures begin with the through-bolts, which do indeed run all the way through the length of the motor. When they are out, you can remove the commutator end of the frame from the field frame. Put the armature in a vise with soft jaws, or place soft material between the jaws and the frame. Remove the shaft nut, pulley, and drive end frame. The brush holders may now be taken out, if new ones are to be installed. It may be necessary to drill out the rivets holding them to the field frame, replacing the rivets later with screws. If anything needs soldering, use rosin flux solder. Examine all the parts, including bearings, brush holder, end frames, and armature. If the mica needs undercutting, follow procedure in Figure 113. If the commutator needs smoothing out, it can be taken to a shop for cutting on a lathe. If it is only dirty, smooth it off with sandpaper and blow out all particles; otherwise, they will cause arcing.

Bearings, armature shaft, loose pole shoe screws, and poor electrical connections can conspire to prevent the motor from turning freely. You can test the purely mechanical aspects simply by turning the armature shaft by hand, once the motor is disconnected from the engine. If it turns freely, there is no mechanical drag. Electrical problems occur both inside the motor and in the system leading to the motor. Basically, that means that—before taking anything apart—all cables and electrical connectors should be examined if any electrical disturbance arises in the cranking system. Next, if generator output is zero, check the ground strap from the voltage regulator to the frame. If it is broken or disconnected, that could cause the trouble. Brushes that stick in the holder can cause generator failure. You can check brush movement by taking a small screwdriver and pushing up on the brush at the commutator side. Either it moves easily (allowing for spring tension on it) or not.

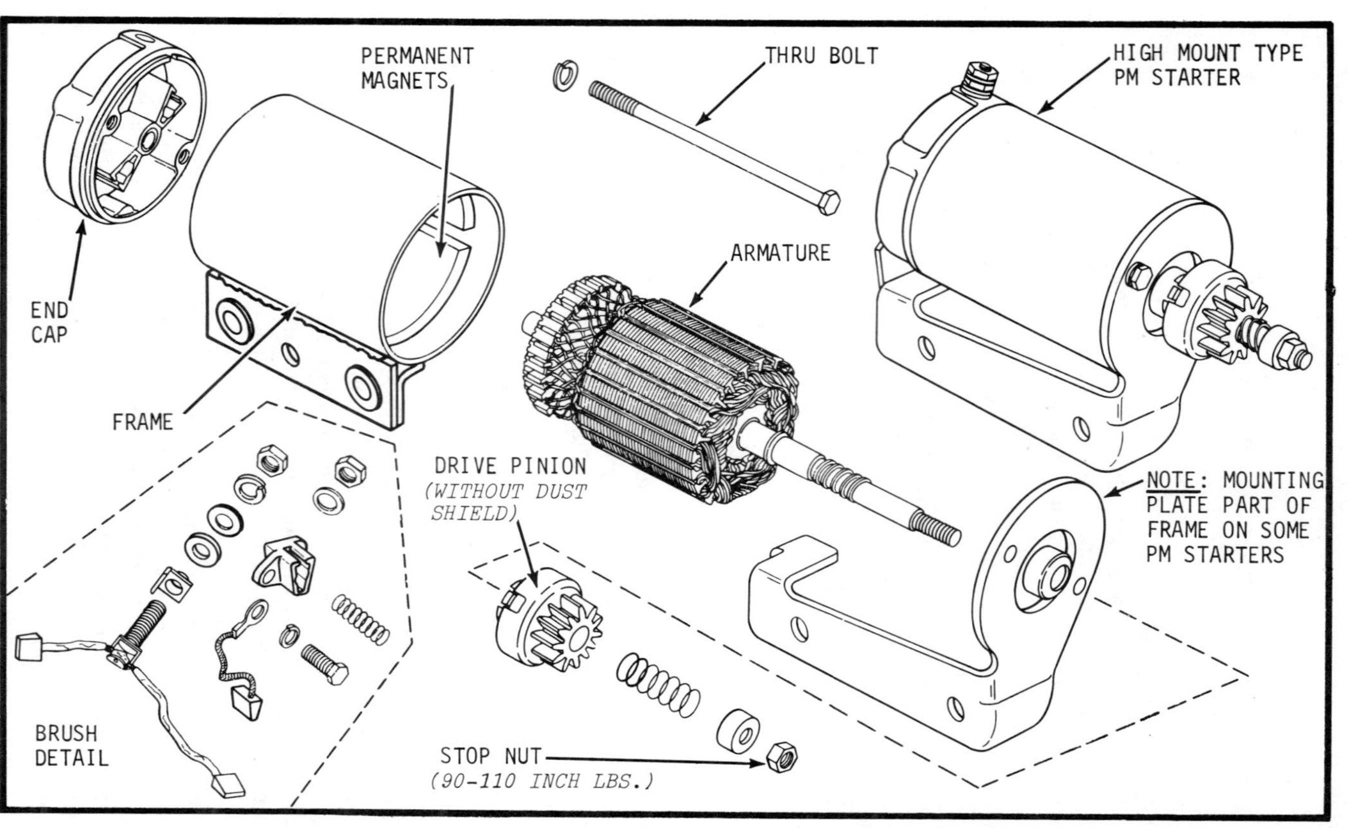

FIGURE 115. Permanent magnet starting motor type in exploded view.

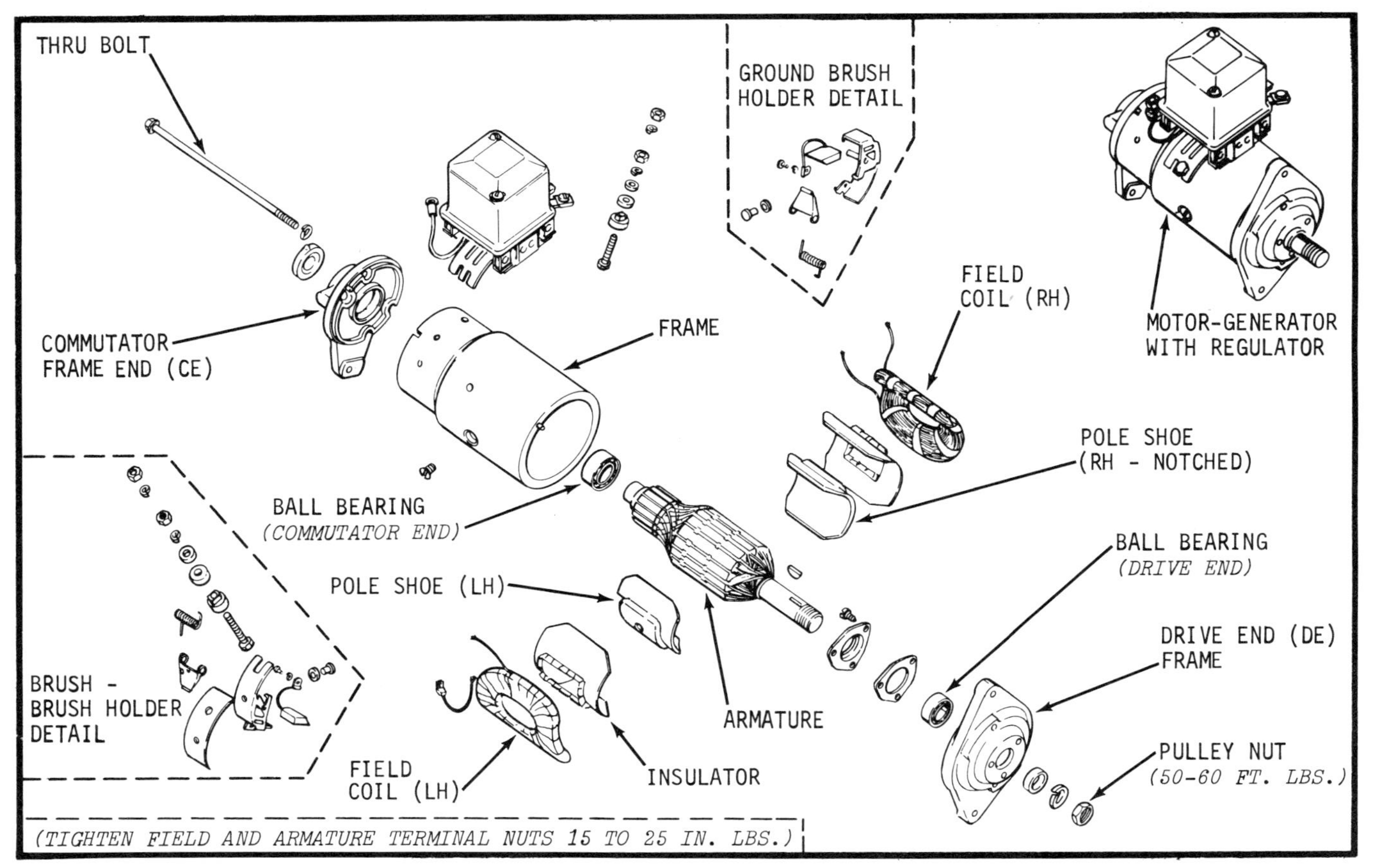

FIGURE 116. Detailed view of Kohler motor-generator with regulator.

If not, you have to do something about it. Clean out the holder and install new brushes. An excessively dirty or gummy commutator can prevent charging or running as a motor. Solder thrown off the commutator indicates that the generator has been overheating, probably from excessive output. This can lead to an open circuit, burned commutator bars, and no output. If everything looks good and you still get no charging, the problem is either a bad connection, bad voltage regulator, or burned-out coil in either field or armature systems.

The voltage regulator, which is more or less identical with old-style automobile types, is a long-suffering beast that rarely goes wrong. However, it does have contact points that sometimes need cleaning and adjusting. The flat point, on top of the coil or armature, develops a small cavity. It should be filed down with a fine-cut riffle file. After filing, wipe off the point completely. Adjust the gap according to specifications in the manual. (It should be .020 inch.) The cutout relay has an air gap and point opening that needs to be checked. Remember: Be sure the battery is disconnected before you do any adjusting on the voltage regulator.

On the cutout relay, to check and adjust the air gap, place fingers on the armature directly above the core and push the armature down until the points close. Measure the air gap between the armature and the center of the core. The gap should be .020 inch. To adjust, raise or

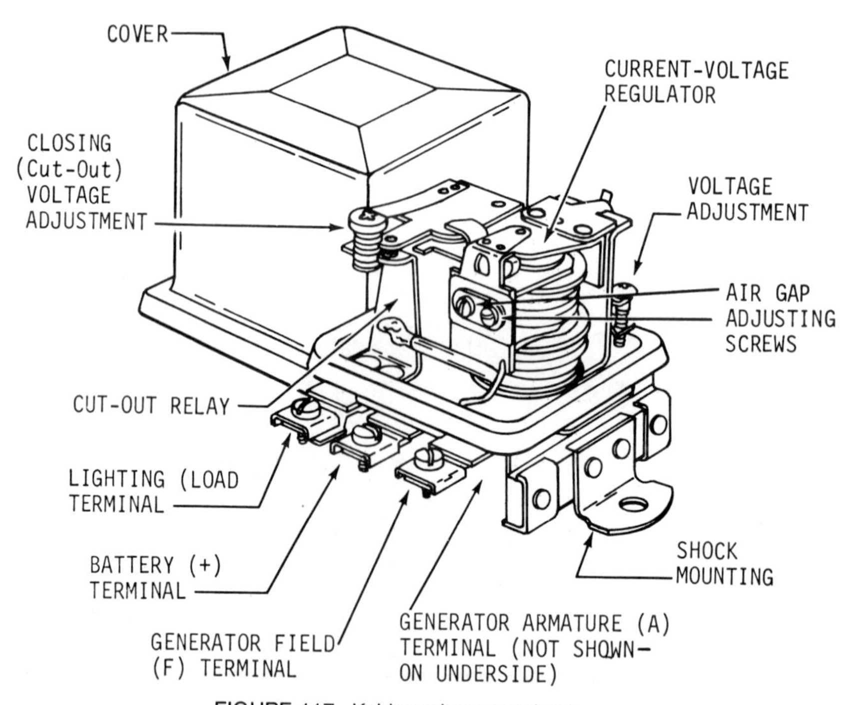

FIGURE 117. Kohler voltage regulator.

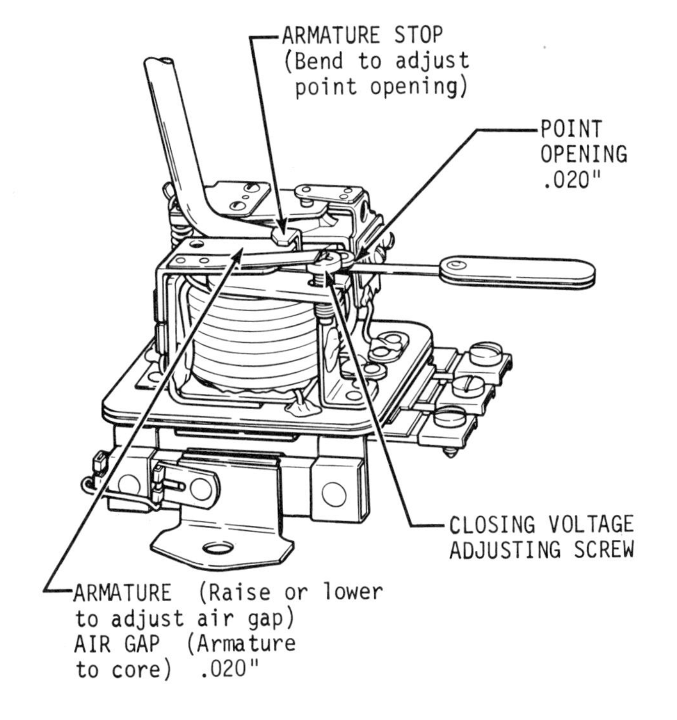

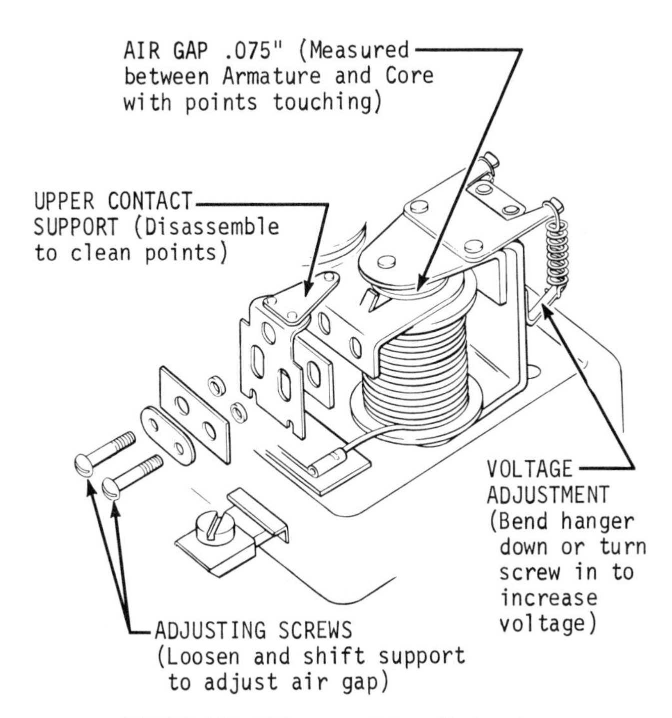

FIGURE 118. Voltage regulator adjustments.

lower the armature at the hinge mounting and tighten the retaining screw after you get the right gap. To adjust the point opening, bend the armature stop until the opening is at .020 inch. You adjust voltage by turning the voltage adjusting screw in to increase or out to decrease the voltage. The procedure requires a voltage meter, so you know what you're doing.

The current voltage unit has two inspections and adjustments: the armature air gap and the voltage setting. The armature air gap should be checked by pushing down on the armature until contact points are touching, yielding a gap of .075 inch. To adjust, loosen the contact mounting screws and raise or lower the bracket to the correct gap. Points must be aligned and screws tightened.

For the voltage setting, which requires the use of a voltmeter, the only inspection is a correct voltage reading. If the reading is between 12 and 14, it is correct. If the voltage is too low, you adjust by turning the adjusting screw clockwise or counterclockwise.

Kohler has a cautionary note: Don't turn the voltage adjusting screw down too far, as the spring may not bounce back when pressure on it is relieved. Incidentally, it is tricky to replace the spring if it loses its oomph. Hook the spring first at the lower end, then stretch it up with a screwdriver blade that you insert between the turns until the upper end of the spring can be hooked.

Kohler alternator charging systems use three different amperage systems—10, 15, and 30 amps. There is no interchangeability of parts.

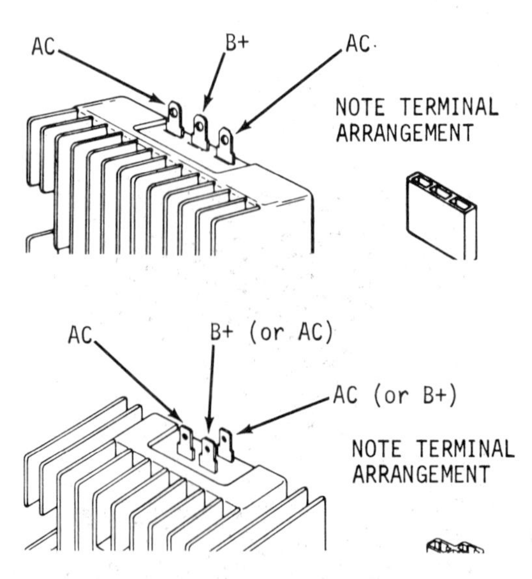

FIGURE 119. Terminals on 10-amp (left) and 15-amp (right) rectifier-regulator units of Kohler engines.

CONDITION: NO CHARGE TO BATTERY	POSSIBLE FAULT/REMEDY
TEST A — Disconnect B+ cable to positive (+) terminal of battery. Connect DC Voltmeter between B+ cable and ground. Check DC voltage; A-1 — If above 14 volts. A-2 — If less than 14 volts (but above 0 volts). A-3 — If 0 volts.	A-1 — Alternator system OK—ammeter may be giving false reading. Repair or replace ammeter. A-2 — Check for defective rectifier-regulator (TEST C). A-3 — Check for defective stator or rectifier-regulator (TEST C).
TEST B — With B+cable reconnected, check B+ (at terminal on rectifier-regulator) to ground with DC Voltmeter. If 13.8 volts or higher, place minimum load of 5 amps * on battery to reduce voltage. B-1 — If charge rate increases. B-2 — If charge rate does not increase.	B-1 — Indicates alternator system OK, battery was fully charged. B-2 — Check for defective stator or rectifier-regulator (TEST C).
TEST C — Unplug leads at rectifier-regulator, connect VOM (multimeter) across AC leads, check AC voltage: C-1 — If less than 20 volts. C-2 — If more than 20 volts.	C-1 — Defective stator, replace with new assembly. C-2 — Defective rectifier-regulator, replace with new unit.
CONDITION: BATTERY CONTINUOUSLY CHARGES AT HIGH RATE	POSSIBLE FAULT/REMEDY
TEST D — Check B+ to ground with DC Voltmeter: D-1 — If over 14.7 volts. D-2 — If under 14.7 volts.	D-1 — Rectifier-regulator not functioning properly. Replace with new unit. D-2 — Alternator system OK. Battery unable to hold charge. Check specific gravity of battery. Replace if necessary.

**Turn lights on if 60 watts or more or simulate load by placing a 2.5 ohm 100-watt resistor across battery terminals.*

FIGURE 120. Troubleshooting a 10-amp alternator system.

CONDITION: NO CHARGE TO BATTERY	POSSIBLE FAULT/REMEDY
TEST A — With B+ cable connected, check B+ (at terminal on rectifier-regulator) to ground with DC Voltmeter. If 13.8 volts or higher, place minimum load of 5 * amps on battery to reduce voltage: A-1 — If charge rate increases. A-2 — If charge rate does not increase.	A-1 — Indicates alternator system OK, battery was fully charged. A-2 — Check for defective stator or rectifier-regulator (TEST B).
TEST B — Unplug leads at rectifier-regulator, connect VOM (multimeter) across AC leads, check AC voltage: B-1 — If less than 28 volts. B-2 — If more than 28 volts.	B-1 — Defective stator, replace with new assembly. B-2 — Defective rectifier-regulator, replace with new unit.
CONDITION: BATTERY CONTINUOUSLY CHARGES AT HIGH RATE	POSSIBLE FAULT/REMEDY
TEST C — Check B+ to ground with DC Voltmeter: C-1 — If over 14.7 volts. C-2 — If under 14.7 volts.	C-1 — Rectifier-regulator not functioning properly. Replace with new unit. C-2 — Alternator system OK. Battery unable to hold charge. Check specific gravity of battery. Replace if necessary.

**Turn lights on if 60 watts or more or simulate load by placing a 2.5 ohm 100-watt resistor across battery terminals.*

FIGURE 121. Troubleshooting a 15-amp alternator system.

Terminals on the 15-amp rectifier-regulator are positioned in a different pattern than those on the 10-amp, for obvious reasons—to prevent the two systems from being joined erroneously. Similar considerations apply to the 30-amp system in relation to the two smaller systems—they aren't interchangeable.

All three systems are more or less foolproof and serviceproof; they are not designed for "field service"—that is, service outside the factory. You can do troubleshooting, as illustrated in Figures 120 and 121. Once you get past the broad outlines of the problems, you are more or less helpless other than to replace major components. Alternators are hardy devices.

You can troubleshoot any of the three systems to determine which component part needs replacement. Troubleshooting differs slightly with each system, as noted in the troubleshooting charts.

In addition to the motor-generator single-unit-type cranking system on Kohler engines, there is also the separate starting motor—charging system type. Chargers can be separate or integrated with the motor. Two types of these motors exist—called PM and Service-Wound Field starters—to go with the alternators.

As to the starting motors that go with these alternator systems, they work about like the starter-generator types discussed, only they don't have the generator abilities. These starters shouldn't be cranked for more than 60 seconds, which is much longer than the time specified for the starter-generator. In other words, they don't overheat as quickly. Bearing, brush, commutator, and drive pinion service are all similar to starting motors in cars. The bearing-brush-commutator service is also similar to the starter-generator motor, which doesn't have the auto-type drive pinion. Drive pinion service is required when the starter doesn't engage the engine even though it turns over, or does so only intermittently.

Sometimes the failure to engage the engine can be cured by putting grease on the spline (but only on the PM-type motor) that pushes the drive pinion out. You can do that without taking the drive pinion off, but you do have to get the spline exposed. There are two types of these splines and drive pinions: one a so-called wound field type, which you do *not* grease; and the PM type, which you do. If this seems hopelessly complicated, it is.

To grease the PM starting motor spline, remove the drive nut, pinion stop, and spring. Clean the drive shaft and pinion and apply Lubriplate Aero grease or its equivalent to the shaft. Install the shield retainer next to the drive pinion, then install the spring and a new pinion stop. Tighten the retaining nut and push the dust shield over the shield retainer.

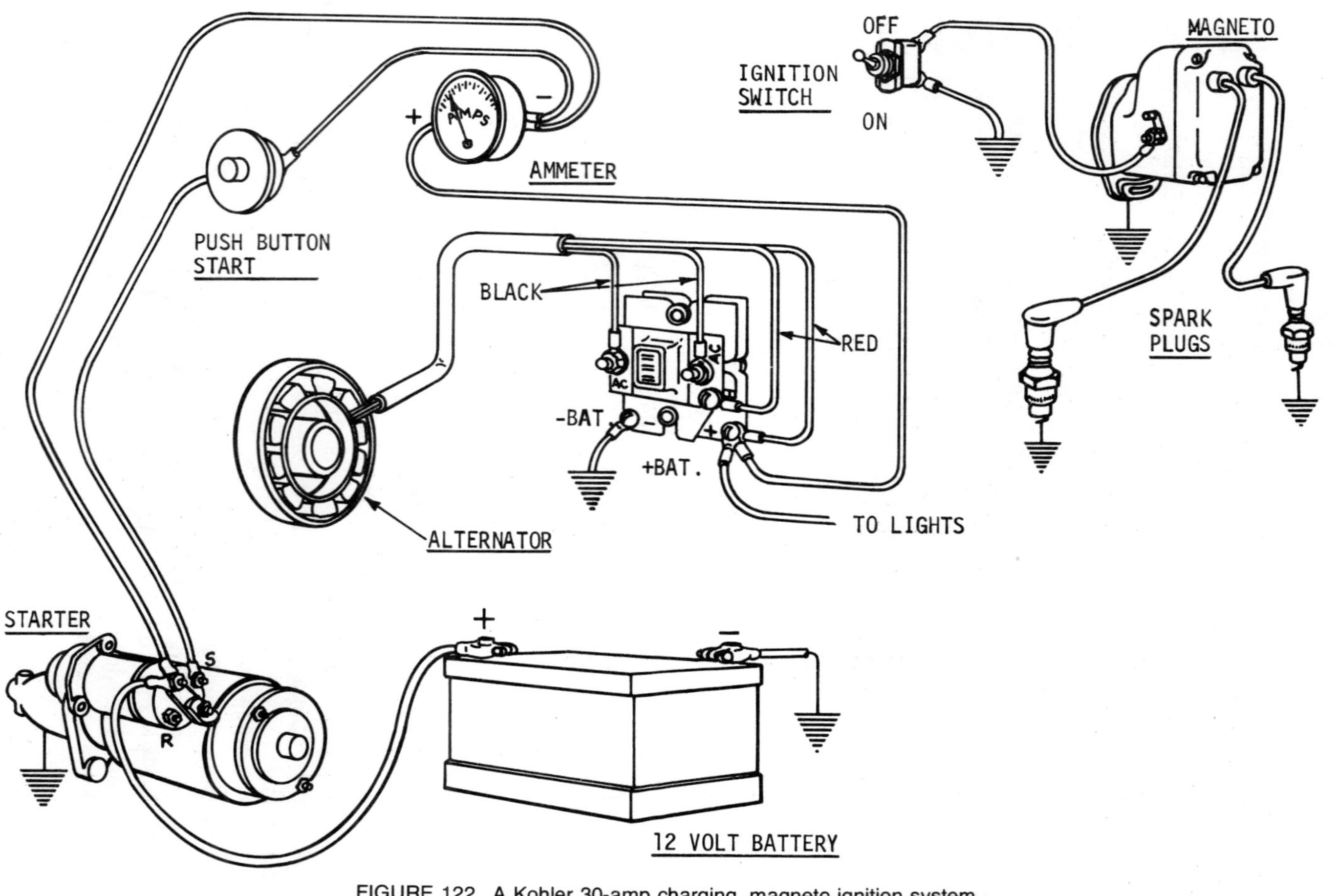

FIGURE 122. A Kohler 30-amp charging, magneto ignition system.

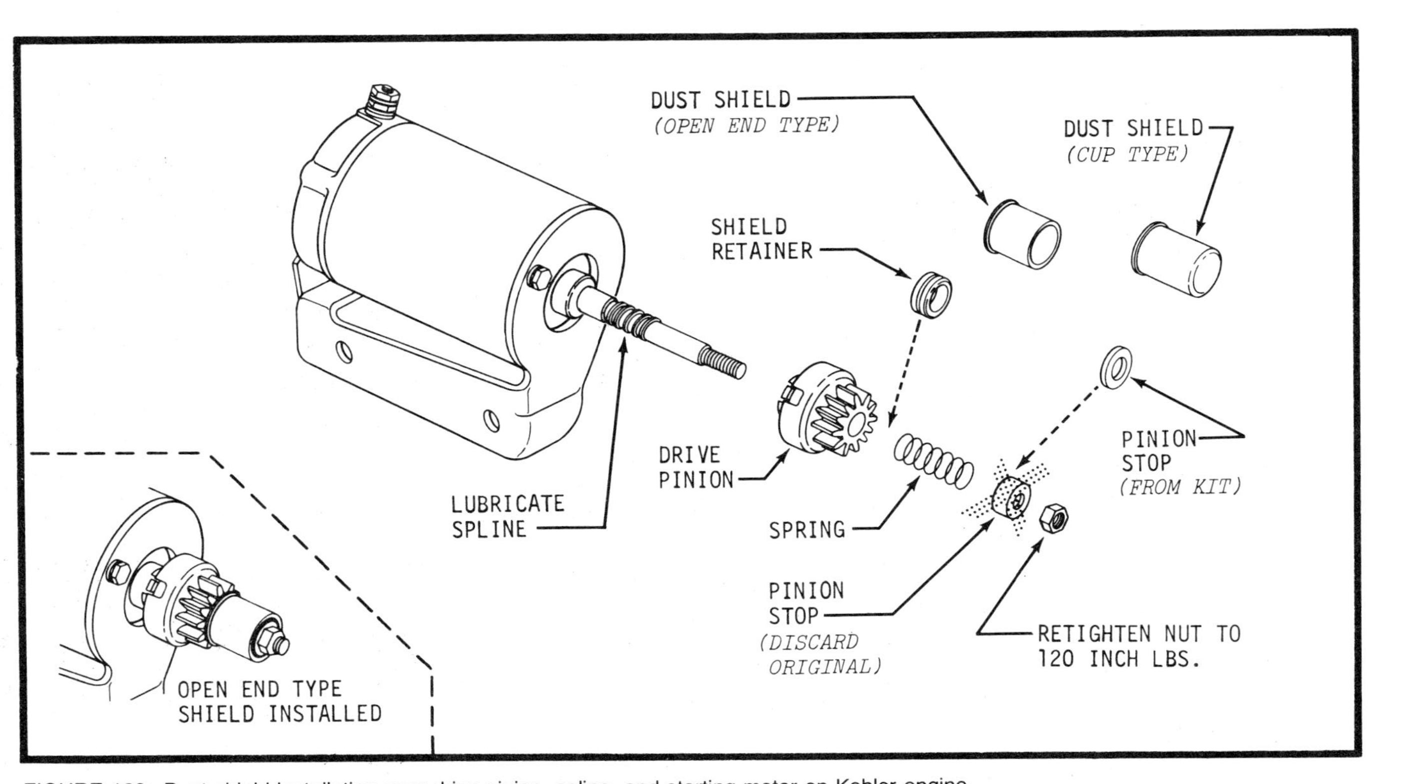

FIGURE 123. Dust shield installation over drive pinion, spline, and starting motor on Kohler engine.

B & S ELECTRIC STARTERS

Starting motors, whether of the automobile type or for small engines, fall into certain patterns of failure. On Briggs & Stratton starters, the following capsulizes these problems. First, the 12- and 120-volt systems:

1. Slow cranking.	a. Added load—engine doesn't turn freely. b. Discharged battery. c. Faulty electrical connections, especially in the battery circuit. d. Dirty, worn starter motor commutator, worn bearings, weak magnets, etc. e. Worn brushes or brush spring. f. Dirty oil or wrong viscosity. g. Extension cord longer than 25 feet (120-volt AC only).

The nicad system (a nickel-cadmium battery) shows the following symptoms:

2. Engine won't crank.	a. Faulty safety interlocks. b. Defective or discharged battery. c. Faulty electrical connections. d. Open circuit in motor starter switch. e. Open circuit in motor. f. Defective rectifier assembly (120-volt only). g. Faulty brushes. h. Faulty solenoid.
3. Starter spins but doesn't crank the engine.	a. Sticking pinion gear—probably dirty. b. Faulty pinion or ring gear. c. Faulty battery. d. Reversed motor polarity—motors rotate counterclockwise viewed from the pinion gear.

4. Starter motor blows fuses (120-volt motor only).	a. Shorted motor switch. b. Shorted rectifier assembly. c. Shorted 120-volt extension cord to motor. d. Armature shorted. e. Circuit overloaded.
5. Starter motor spins, won't stop.	a. Defective starter switch.

The nickel-cadmium system in series 92000 and 110900 engines has a motor and switch, wiring harness, rechargeable nickel-cadmium battery and battery charger. The ignition key, when turned on, permits the battery to send power to the motor that cranks the engine. After about 40 to 60 starts the battery needs recharging. This is accomplished through the charger that is plugged into a 120-volt household outlet. It takes 14 to 16 hours, and it doesn't work well below 40° F.

In every case when a starter won't work it is wise to make sure that the engine turns over properly. Remove the spark plug and turn the crankshaft over by hand. It should turn readily.

One begins troubleshooting the system by checking the battery. If you have reason to believe the battery is defective, the best thing to do is to have it tested at a shop.

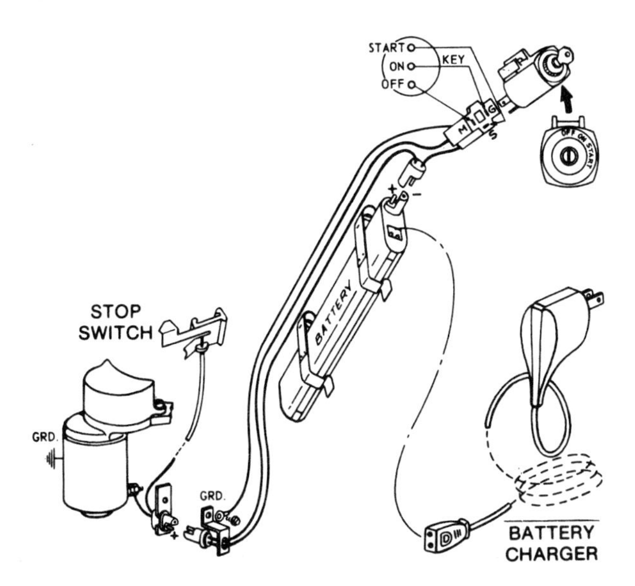

FIGURE 124. Briggs & Stratton wiring diagram of charging and starting system on series 92000 and 110900 engines.

The motor drive and clutch can be tested: when the starter switch is turned on, the nylon spur gear should rise to engage the flywheel ring gear and then crank the engine. You can see whether this is happening by removing the starter cover. If the starter motor drive doesn't react properly, examine the helix and the nylon spur gear for binding; the spur gear must move freely on the helix. Don't oil the gear or clutch helix; if they are sticking, they will need cleaning out or replacing.

The starter motor clutch prevents damage from shock such as engine backfire. It should not slip in normal engine cranking. If it does, you need a new one. Don't replace it if it continues to crank the engine, no matter how much it slips. If replacement becomes necessary—a statistically rare event—it requires the removal of the starter cover, spur gear retainer, and the starter gear cover and gasket, all of which are above the clutch. Three screws hold the gear cover and gear. Once they come out, the clutch assembly can be lifted out with the pinion gear.

If your starter motor falls prey to several of the symptoms listed in the charts, it may need more than peripheral work. Disassembly of a nickel-cadmium starter motor isn't difficult, though it can be time-consuming. You have to weigh the time and what the shop will estimate the bill might be.

You take off, in order, the starter cover, nylon spur gear retainer, and the nylon spur gear. Three screws hold the cover and gear; remove them. Lift out the clutch assembly and the pinion gear from their shafts. Next to fall are the through-bolts; then the motor end head comes out of the housing. Push out the armature through the bottom of the housing, sliding out the rubber mounted terminal along with the end cap. Check the brushes before removing the armature to see how freely the brushes are moving. If they are sticking, that's a basic troublemaker. If they are worn down to a length of ¼ inch, they must be replaced. When you do that you must swab out the brush housings, using whatever cleaning method appeals to you, including an old toothbrush dipped in some solvent such as carburetor cleaner. If the brush cavities are corroded, it will take more than a brush; you'll need to scrape. Fingernail files work here. Brushes get pushed by the armature constantly as the motor turns, and they are made of soft carbon that rubs off into corrosion, hence the need for cleaning.

The armature is a more delicate customer. Clean every speck of dirt from it that you can see—the end cap, motor support, the gears nearby. The armature commutator—that copper segmented, circular thing—should be cleaned with fine sandpaper. Don't use aluminum oxide paper or emery cloth—their dust can cause arcing and brush ruination. If you have any reason to suspect a defective armature, you must take it to a shop for testing in a "growler." It does no good to use a voltmeter or ohmmeter or whatever; they aren't sensitive enough for the task.

FIGURE 125. Exploded view of Briggs & Stratton starter.

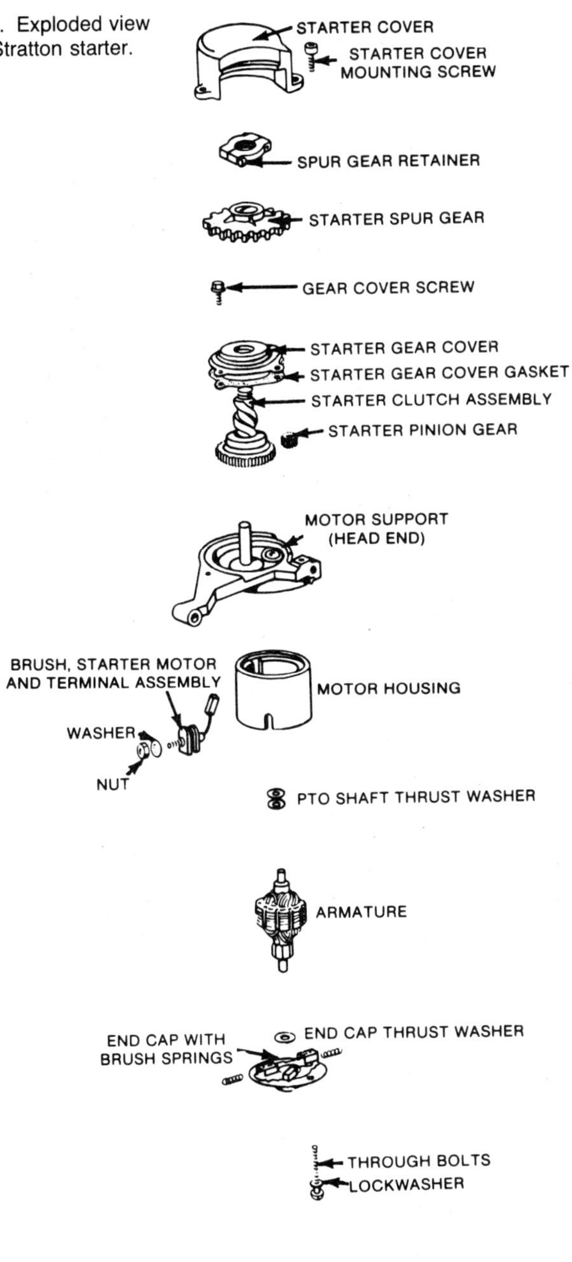

That's about it, as far as disassembly of a starting motor goes. Field coils can be tested for continuity with a continuity tester of some sort, or with an ohmmeter. In general, if an armature has short-circuits, they will show up as sluggish performance—too slow—or as high voltage in a voltmeter test; current draw will be well above 12 volts, using a voltmeter.

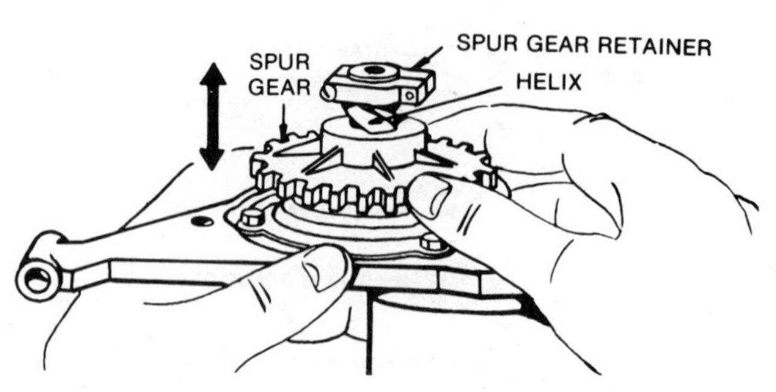

FIGURE 126. Spur gear and helix on the starter drive.

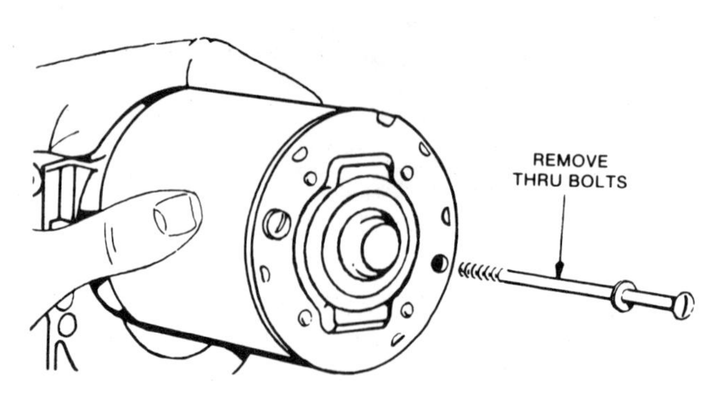

FIGURE 127. Removing through-bolts on starter.

To put the starting motor back together is the reverse of what you have just done, allowing for cleaning and the lubrication of bearings. Replace brush springs and brushes in the holders, holding them in position with some kind of restraining device. Thrust washers go on the armature shaft. Slide the armature shaft into the end cap bearing. If your brush holding device doesn't work, you will find it out quickly at this point—the armature will bang into the brushes. That's a no-no. If you can't figure out how to get a holding device for the brushes in place and working, find a neighbor who has nothing else to do and is willing to lend a couple of hands. Next, replace the rubber mounted terminal, put the remaining thrust washers on the other end of the motor shaft, install the end head over, and the through-bolts. Notches in the end cap, housing, and end head must align. Now check the armature to see that it moves freely. Next, the pinion and clutch gears go on their shafts after you grease all the gears. Install the gear cover and gasket. Tap the end cap edge a bit with a soft hammer to align the bearings. Put back the nylon spur gear and retainer assembly and tighten the screws. Put back the starter cover. Now the starter motor is ready to be installed on the engine, exactly as you removed it—in reverse. There are no tricky adjustments.

Nickel-cadmium batteries on the Briggs & Stratton nicad system are usually strapped to the lawn mower handle, with cables running up to the ignition switch and down to the starter. Whenever the battery position is changed, for whatever reason, it is important to allow the correct amount of slack in the cables so that they don't bind at the bottom connection with the starter.

If you are installing a new battery, follow directions—don't assume that you can't make a mistake in so easy an operation. Beware of overtightening the battery holder; also, connect the *positive lead first,* to eliminate the possibility of sparks from an accidental *grounding.*

In Briggs & Stratton models larger than those discussed, the motor and drive differ somewhat. For example, model 130000 has a permanent magnet motor and different location for the starting gear drive, though details don't differ that much. Both 12- and 120-volt motors are found on these larger models. They are similar enough to be discussed together.

The gear system and motor come apart after you remove the locknut. Then you can get out the helix, drive gear assembly, and pinion gear for cleaning if that is the problem. Don't do anything before consulting the troubleshooting charts shown in Figures 120 and 121. If there is any failure in the starter system, either mechanical or electrical, you should try to pinpoint the problem before undertaking disassembly. If the motor doesn't crank, the fault can be in the engine (not likely) or in

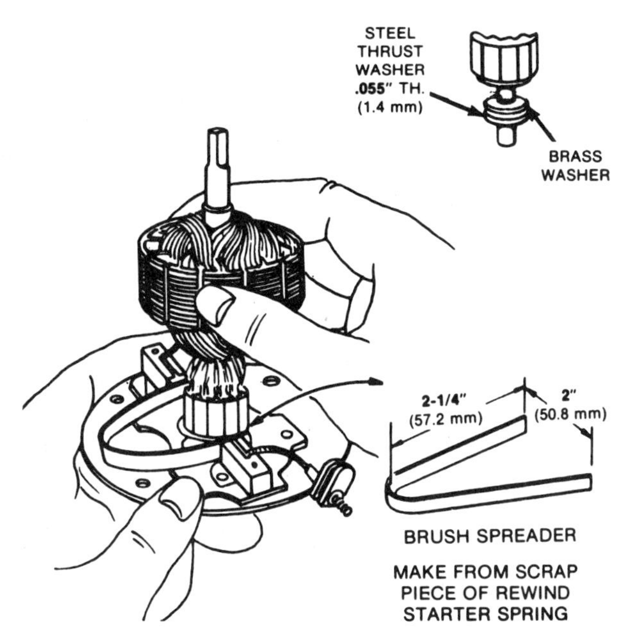

FIGURE 128. Getting the armature back in its housing, with a brush spreader.

the gearing (probably). It can also be in the starter motor. Gearing problems usually start with dirt and corrosion, and cleaning thoroughly will solve them. If cleaning doesn't work, you have to buy new parts, and usually the design of these things is such that you have to replace whole assemblies rather than single small parts. That's the way of the world as well as of starter motors.

The 12- and 120-volt starter apparatus can be checked visually by removing the starter and turning it over to see whether the pinion gear rises to engage the flywheel ring gear. If it sticks (doesn't rise), take it off and clean it thoroughly, but don't lubricate the drive assembly. That

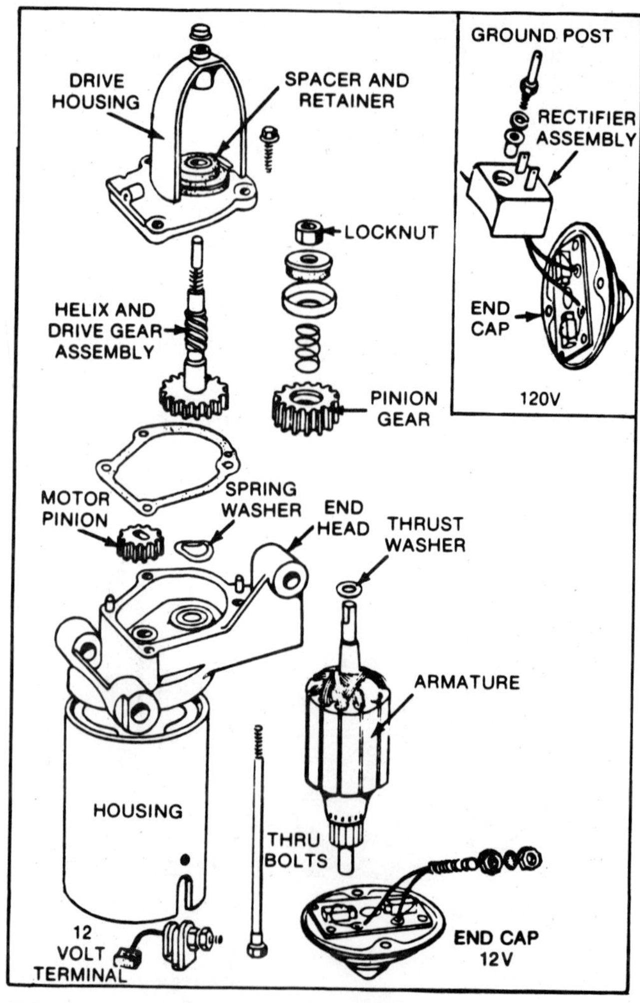

FIGURE 129. The 12-volt and 120-volt starters, in an exploded view. The motors are the same, but current flow differs.

will merely make it stick all over again. You must also inspect it for wear, cracked gears, and rough spots. Defects of this sort require replacement of the assembly.

Note that whenever a ring gear must be replaced—which should never happen, but it can—(it's on the flywheel and is what gets turned by the starter pinion gear)—a steel ring gear must be used if the pinion gear is made of steel. An aluminum ring gear must be used if the pinion gear is nylon. Rings are usually held with rivets that must be drilled out and replaced by screws. Use a 3/16-inch drill and clean out the holes after drilling.

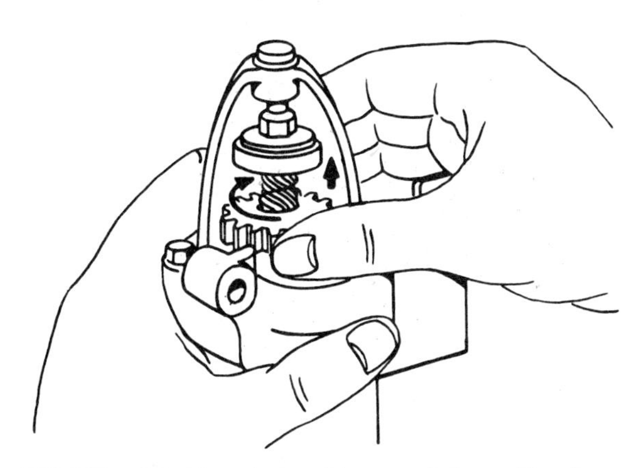

FIGURE 130. The starter drive must be guilty of insurrection—it must rise up—to do its duty.

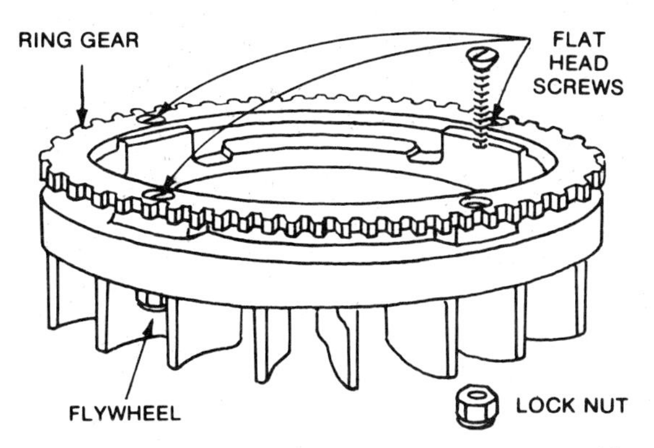

FIGURE 131. Replacement ring gear goes on like this.

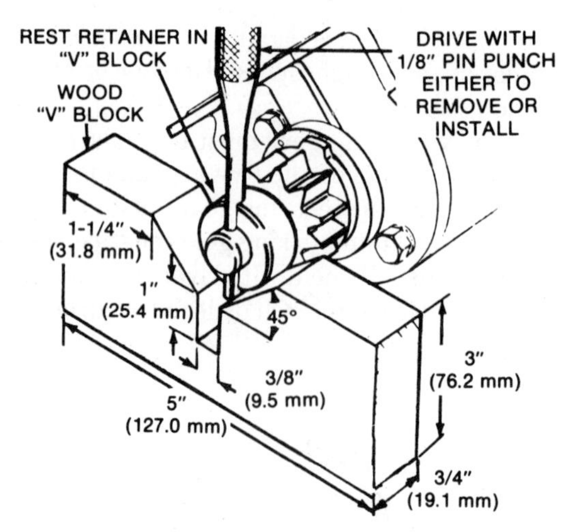

FIGURE 132. Driving out roll pin with punch.

Disassembly and assembly vary from model to model. In some of these starter assemblies you have to drive out a roll pin with a punch to remove the gear assembly retainer. Mostly you get at them by removing a locknut.

If yours does have a roll pin, it may also contain a spring cap assembly—a plastic cap over a gear return spring. The cap snaps off with two opposing screwdrivers to pry it off. It goes back by putting a $^{13}/_{16}$-inch socket over it and tapping lightly on the socket. No mystery.

Note that 120-volt alternating current motors, which plug into your garage wall outlet for starting, have rectifiers that change the alternating current to direct current (AC to DC). If you can't figure out any other reason why the motor won't work, consider the rectifier as the culprit of last resort. The rectifier itself consists of a switch, an AC three-wire ground receptacle, and four rectifiers in an epoxy case. Tests require an AC voltmeter and a DC voltmeter. You'll have to take it to a shop. Rectifiers can't last forever; they can quit instantly, without warning, though in alternators they may make telltale noises.

Briggs & Stratton starter generators resemble those by Kohler, discussed earlier. These devices have regulators and batteries. They attach to the engine with two belts and pulleys.

The point of having a starter-generator is to keep a battery charged up constantly, as in a car, and to combine two essential functions into one component. However, the battery must be nearby to make use of the device—on a riding mower or other larger power equipment.

The 12-volt belt-drive starter automatically engages a belt clutch when it is turned on. The belt clutch disengages automatically when the engine catches, but before turning on this starter, the driven equipment

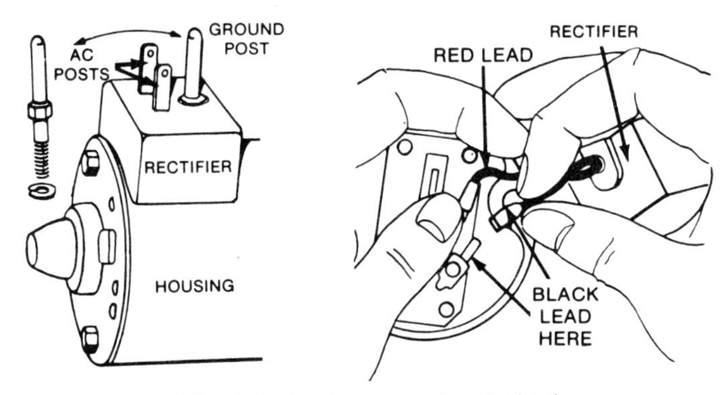

FIGURE 133. Rectifier assembly and leads.

has to be disengaged from the engine. The starting system can handle only the engine. It's like putting the car in neutral before turning on the ignition.

Belt adjustment and/or replacement is about what you'd expect. If you've adjusted or changed one engine belt (or even a belt on a furnace, dryer, etc.), you've done them all—or almost all (there's always the Dodge Omni belt).

Briggs & Stratton uses two different versions of belt adjusting details. In Figure 134, the nuts "A" and "B" are to be loosened slightly. Then move the starter motor away from the engine as far as possible. Rock the engine pulley back and forth and at the same time, and slowly, slide the starter motor toward the engine until the starter motor pulley stops being driven by the V-belt. Move the starter motor another 1/16 inch toward the engine. Then tighten nuts "A" and "B" as before.

Alternators become an issue with the biggest of the small engines. An alternator is a separate generator, used to generate the voltage required to keep the 12-volt battery charged. The alternator, unlike the old generator, produces alternating current, which it changes to direct current by means of a system of diodes or rectifiers (the generator produced direct current).

Alternators on the model series 130000 are rated at 1½ amps at 3,600 r.p.m., less at lower speeds. This alternator does not have a voltage regulator. If the battery is run down, check battery polarity. Negative should be grounded to the engine or frame, positive to the starter motor and alternator charge cable. If it gets reversed, the rectifier can be ruined. To test for output, disconnect the charging lead from the charging terminal; be careful to prevent the charging lead from touching the engine or any part of the equipment. Clip a 12-volt testing light between the charging terminal and ground. Start the engine. If the testing light goes on, the alternator is working. If not, it is not. (You can use a 12-volt automobile lamp.)

You can test the stator (stationary coils) on the alternator by

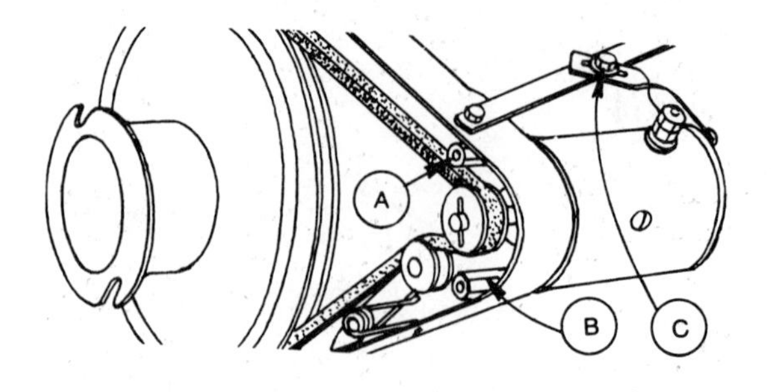

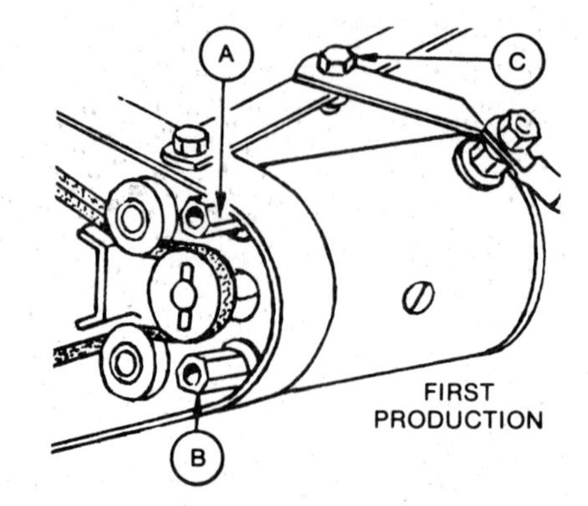

FIGURE 134. Two belt-tightening sequences on starters.

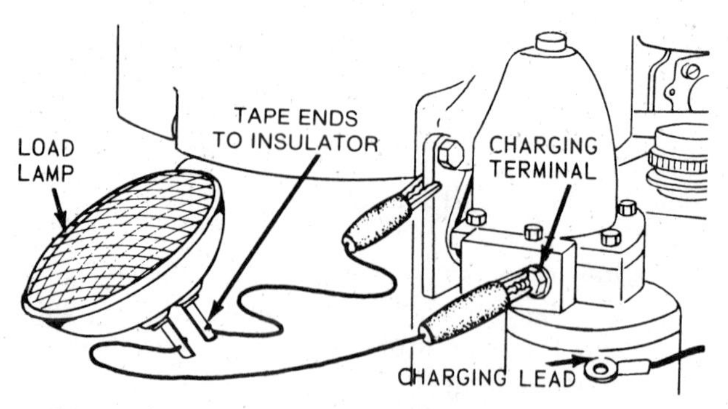

FIGURE 135. Testing an alternator with a 12-volt light.

unplugging the connector or charging lead from the battery and rectifier assembly. Remove the screw from the ground wire or rectifier assembly from the starter motor. Turn the rectifier assembly to expose wires attached to soldered terminals. Ground wire or rectifier assembly must not touch the engine in the test. Start the engine, and while it runs pierce stator wires with probes from a 12-volt light (or touch the terminals in the rectifier box).

If the light does not go on, the flywheel magnet or the alternator stator is not working. First, eliminate the flywheel magnet. Does metal stick to it? That doesn't settle the matter conclusively, but it's a start. One should accept the fact that magnets expire much less frequently than coils. If the stator fails the light test the probability is that the stator is at fault rather than the flywheel magnet.

It's a pretty big job to replace a failed stator in an alternator, especially if you are skeptical of the test. One trouble with such tests is that they are rarely as conclusive as one would prefer, given the fact that much work is involved that could fail to rectify the matter. The alternative is, of course, to take the motor (alternator, etc.) to a shop. You know what that means. Contrary opinion in this matter always goes to the effect that no matter how many parts you replace on your own needlessly, before hitting the right one, you are always ahead of the game. Labor costs at a shop, plus the retail costs of parts, are sufficiently generous as to cover a lot of your own bungling. Bungling is a luxury, but even if you do replace parts that don't need it at this time, you *do* have a new part.

To replace a stator in the alternator, remove blower housing rotating screen, clutch assembly, and flywheel. You must remember or write down the location of stator wires—those under one coil spool and those between the stator and drive unit housing. Remove the ground wire or rectifier assembly from the starter drive housing and take out the two stator mounting screws and bushings.

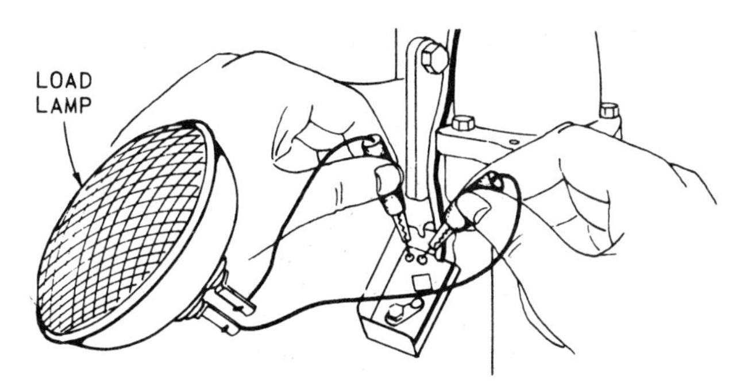

FIGURE 136. Stationary coils on an alternator can be tested with a lamp and prods.

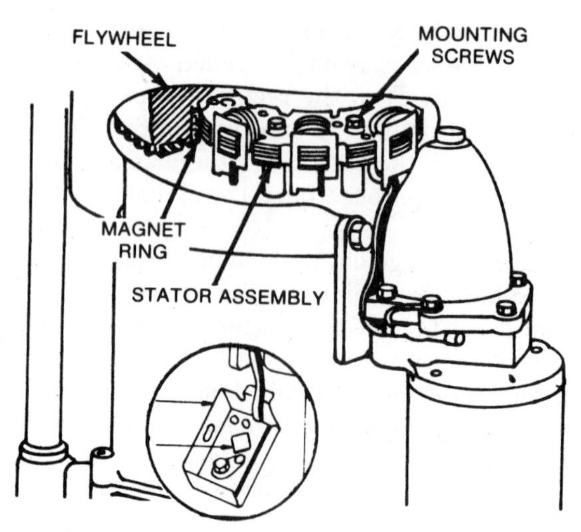

FIGURE 137. Replacing a failed stator in an alternator. Note mounting screws and relation of stator to flywheel and starter.

The new stator assembly must match the old one and go back in precisely, even as to the way the wires are positioned once they are connected. Tighten mounting screws as you push the stator toward the engine crankshaft to take up clearance in the bushing. Before reassembly, push the stator wires against the cylinder so as to clear the ring gear and the flywheel. Attach ground wire (or rectifier assembly—models differ as to cable attaching details) to the drive housing, and replace flywheel and clutch housing, screen and blower housing.

You can test the rectifier (sometimes called diodes) using any voltmeter and/or ammeter. The test is done with the engine *not* running. Test resistance from the charging terminal to the ground (frame, deck, etc.). One way there should be a meter reading; the other way there should be no reading. The reading itself does not matter; you look for a flick up of the needle. If the meter shows a reading both ways or

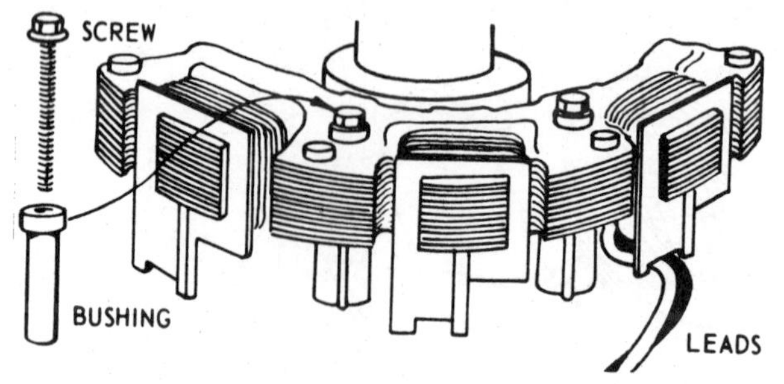

FIGURE 138. In assembling the new stator, leads and other connectors must go back exactly as taken out.

neither way, the rectifier must be replaced. You have to cut the stator wires close to the rectifier so that they remain as long as possible. You must then twist and solder each stator wire to the new rectifier wires and insulate them with tape. Also, remove and throw out the original ground wire from the drive housing and replace it.

Briggs & Stratton also makes an AC-only alternator—one that is used without a battery, to power lights on outdoor equipment. Needless to say, these are flickering lights; as engine speed (hence alternator) increases or decreases, the lights follow suit, becoming now dim, now bright.

What we have done is to repair an alternator, one that is less complex than its similar unit on the family jalopy, but more or less identical with it. The pitfalls involve soldering, wiring, and correct assembly. If you undertake such a job, thinking thereby to do a similar number on the jalopy alternator, the fact is that you can buy a rebuilt car alternator for about $35 or $40, whereas the similarly rebuilt Briggs & Stratton alternator would not be readily available, unless you happen to live near a place that does such work. They are not all that common. Alternator rebuilding is a growth industry only within severely limited parameters—to use up my jargon quotient for the day.

Briggs & Stratton also has a dual circuit alternator. It operates as an integral part of the engine, separate from the starting system. In effect, it is two separate charging systems. A single ring of magnets supplies magnetic field for both. One alternator uses a solid-state rectifier and charges the battery. The other system keeps the lights on. Use of the lights doesn't drain charging ability; electrically, the systems are independent. This hydra-headed alternator is best left to experts

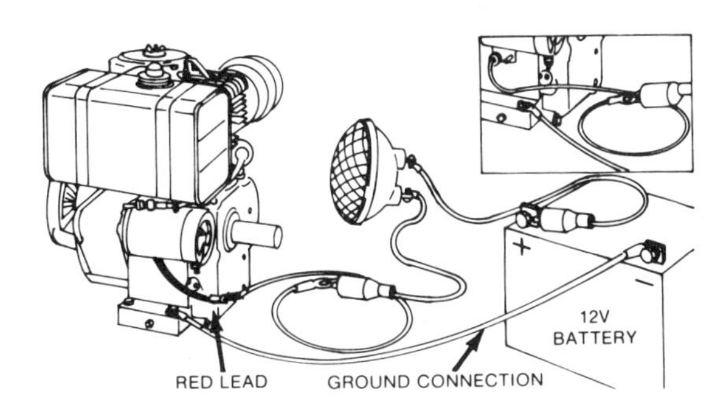

FIGURE 139. To test for short in diode (rectifier), you need an ohmmeter or voltmeter or a test lamp connected as shown here. With a lamp, disconnect charging lead from battery and connect lamp between battery and lead. (Engine running.) If lamp goes on, diode is defective. Disconnect charging lead at alternator. If light does not go out, lead from cap to battery is shorted. If light goes out, diode is defective. Connector assembly must also be replaced.

when trouble develops; however, many of the same tests we used earlier can be applied to both heads.

Diodes can be tested in exactly similar manner; disconnect the connector plug and use an ohmmeter. Attach one meter lead to the red stator lead wire (which may require piercing insulation), the other to the diode lead in the plastic connector (which you find by tracing via color code). The meter should read in only one direction; if it shows a reading (no matter what) in both directions or does not show a reading in either direction, the diode (rectifier) must be replaced, and the connector assembly along with it.

The stator charging coils can be tested, visually at first (and that is always the first test in any electrical troubleshooting), looking for insulation and cable breaks and burns, burned spots indicating short-circuits, etc. To do it correctly in the present case, you have to remove the blower housing, flywheel, and other paraphernalia and look especially at the red lead wire. If you have an ohmmeter, touch one lead to the stator (stationary field coil) laminations, the other meter lead to the red stator charge lead wire. You should get a continuity reading. Then remove the screw that attaches the stator ground wires to the stator laminations. Ground wire must not touch laminations in the test; repeat the previous test. Now the meter should *not* show continuity. If it does, replace the stator.

A discoloration of stator coils is not necessarily a sign of a defective stator. A shorted diode can cause this condition by passing battery current through the stator coils to the ground. The resulting heat causes the discoloration. A new diode will cure this problem. That's for the shop—or the venturesome.

The dual circuit type alternator has various styles and models, depending on the equipment and load. All may be checked and diagnosed along lines discussed above, but some of them require specialized test equipment you won't have. However, any ohmmeter (and a combination volt- and ohmmeter) can take you far. Every house ought to have such devices or a single volt- and ammeter combined, and every housewife should know how to use them.

The 7-amp alternator, used by Briggs & Stratton, is yet another in this alternator group, found on model series 140000, 170000, and 190000. This alternator uses a rectifier and a solid-state electronic regulator for rapid battery charging and/or extra electrical loads. The function of the regulator is to protect the battery from overcharging. An unusual feature of this 7-amp alternator is something called an isolation diode assembly, which is a kind of electronic check valve. It backs up the regulator, which might not keep the battery fully charged under all operating conditions. It is not an essential device. Equipment manufacturers may elect to eliminate it by routing the alternator output lead through a special ignition switch that disconnects the alternator when the switch is in the "off" position. It can also be installed as after-market equipment on engines that don't have it.

The isolation diode can conk out, requiring fairly simple testing. A fuse and holder should be disconnected to connect a test light between the tip of white wire and the battery negative terminal. The light should *not* go on. If it does, the isolation diode must be replaced.

The Briggs & Stratton 4-amp alternator, without a regulator, can be tested much as with the alternators discussed above. It has a solid-state rectifier and a fuse. The rectifier is attached to the blower housing baffle in a small black box. If the alternator fails to charge the battery, follow the same testing procedure as discussed earlier. Test the stator and rectifier for short-circuits; the stator for visible wiring breaks, bare leads, and other visible defects; test the stator coils for continuity; and test the rectifier diode. A lead from the black box connects to a single pin in the detachable plug. With the box in place on the blower housing, test the rectifier with an ohmmeter, from the pin to the blower housing. Then reverse meter leads and check again. The meter should show a reading in one direction only. If the rectifier pin shows a meter reading both ways, or shows no reading either way, the rectifier must be replaced. It means the diode is shorted out.

There are two other alternators in the Briggs & Stratton arsenal, including a tri-circuit and a 10-amp alternator. Though some testing details differ, they follow patterns similar to those discussed.

TECUMSEH ELECTRIC STARTERS

Tecumseh offers a troubleshooting chart for its line of electric starters and supporting paraphernalia—alternators, regulators, etc. Though the following diagnostic procedures are for the four-cycle engine primarily, they apply equally to the two-cycle.

Once past the troubleshooting stage—largely an intellectual exercise in eliminating problem areas—Tecumseh recommends these preliminary tests: check battery charge (but it can be done only if you have test equipment; otherwise, you cart it to a shop); ascertain that the switch and solenoid between battery and starting motor are working. This you can test, using a light or any continuity tester, including an ohmmeter. Examine terminals; they must be bright and clean if you expect unimpeded electrical flow. Examine all wires and cables inch by inch for insulation and other breaks. If the starter does engage the flywheel but can't turn the engine over, examine clutches, pulleys, and the engine itself for excessive friction.

In making electrical tests, separate testing of copper components from iron. Copper includes wiring in armature and field coils as well as the brushes and commutator bars. Iron includes case and cover, armature shaft, and the laminations. Needless to say, check copper first; it's soft, conducts the most destructive part of the electricity, and (ironically) gets all the wear.

To test for grounded starter, use some kind of continuity tester, including an electric light and probes. One probe goes to starter housing, the other probe to the field terminal post. If the light goes on, the copper portion of the starter is grounded. But don't replace the motor yet; continue on. Make all the tests. Remove brushes from their holders. Put one probe on the brush, the other on the terminal post. If the light does not go on, the field coil is open (ruined).

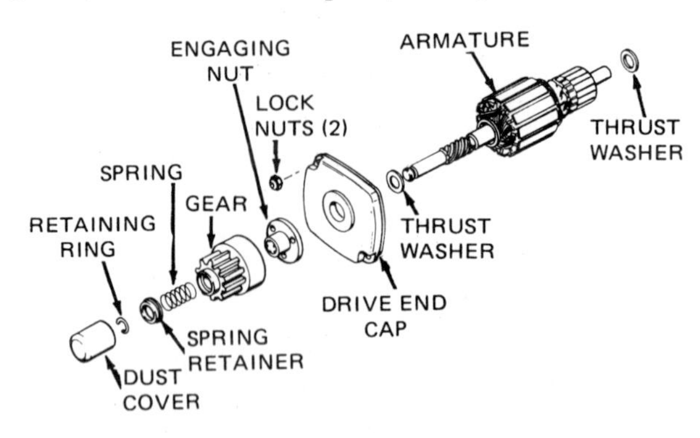

FIGURE 140. A Tecumseh starter showing all major parts.

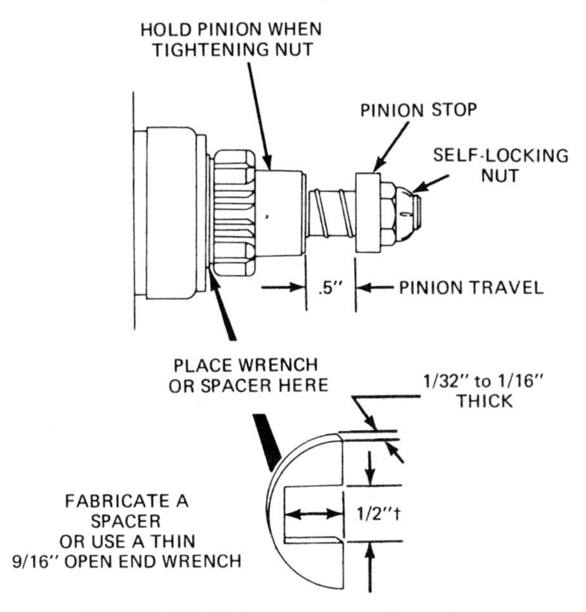

FIGURE 141. Pinion gear assembly end.

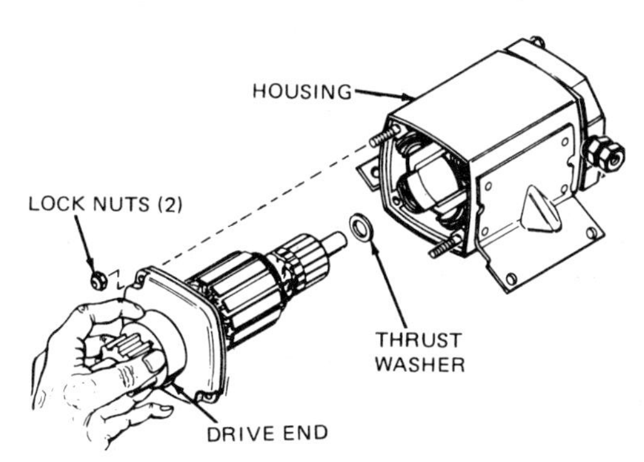

FIGURE 142. Drive end, rotor and commutator, and housing of Tecumseh starter.

Visually check the armature for burned winding insulation, and check the commutator bars for dirt, oil, grease, or other problems. The grooves between the bars must not be clogged. An open winding can cause the trailing edges of the bars to be burned. What happens is that a break in the circuit (open winding) that contains current flow will cause extremely high resistance in that circuit. The open circuit winding will be attached to the bar whose trailing edge is badly burned. All the loops are connected in series, and when an open circuit occurs it will reduce the current path in the armature by one-half. Thus, the brushes are breaking the circuit twice for every revolution of the armature; the resultant arcing causes burning of the brushes and commutator bars.

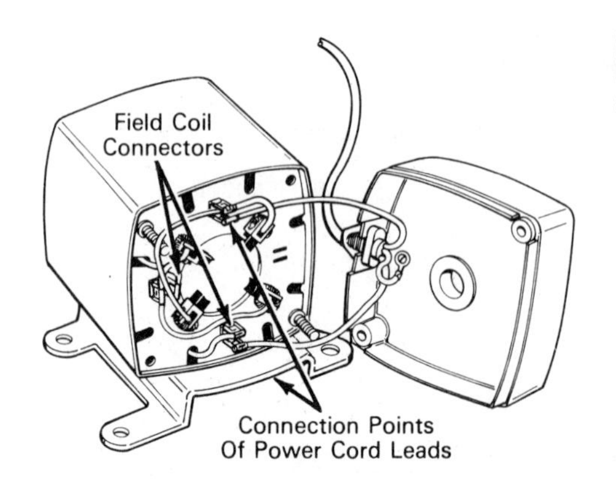

FIGURE 143.
Electrical connectors
on Tecumseh starter.

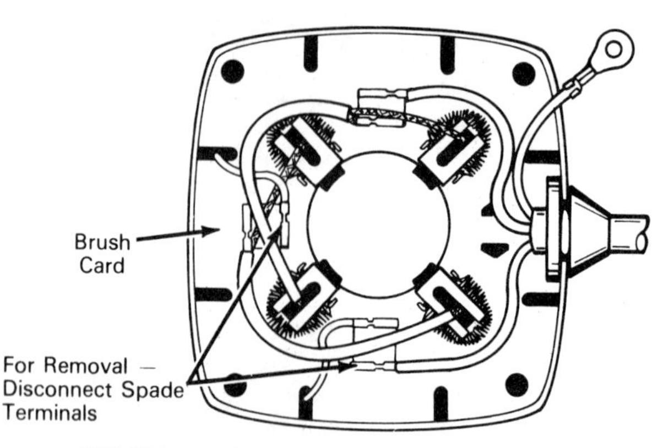

FIGURE 144. Brush holders fit into "brush card."

If the commutator is only glazed and/or dirty from prolonged, normal wear, it can be cleaned and restored. Use 00 sandpaper (not emery paper) and turn the commutator (ideally in a lathe, but hand-turning will work unless the commutator is in need of filing off). After turning down the commutator bars, or sanding them off, the mica insulators between the bars must be cut down to a depth equal to the mica width. Mica undercutting tools are available and inexpensive. Cut particles must be blown out, preferably with compressed air. If you don't have a compressor or the use of one, use a brush and your own breath, going over the territory repeatedly.

The fact is, the only true test of an armature is what the tester called a "growler" provides. So, if there is any doubt in your mind, take the armature for that test to a shop. They'll do it inexpensively or perhaps even without cost.

Tecumseh makes 10- and 20-amp alternator charging systems that have similar but not interchangeable electrical components. Their testing can be done along lines discussed earlier, but you must have a reliable direct current voltmeter and a tachometer in order to measure engine speed.

ENGINE REPAIRS

Small engines can run on forever without requiring much engine overhaul. If you change oil regularly, and don't make the engine do tasks it isn't equipped for, you shouldn't have to dig into the inner engine. Nevertheless, anything that can wear out will, and many things that don't get much damaging wear can self-destruct. Every moving part inside an engine, if properly oiled, can endure for many years, but each part may require replacement a week after you get the thing home. Metals and engine parts are not predictable and it is easy to abuse a small engine—a lawnmower blade can hit a rock or other object disastrously; a snowblower can encounter something in the snow that it cannot digest. Tillers, riding mowers, and other gasoline engine equipment can suffer unexpected blows. What follows includes major engine overhaul. It should not be confused with similar work on a car engine. For one thing, you deal with a fourth to an eighth the number of parts; disassembly is incomparably easier, and complexities and frustrations are almost nonexistent, whereas auto engine overhaul is loaded with them every step of the way. That does not mean that everything is a "piece of cake." It isn't, as we will see.

Perhaps the best thing about working on a small engine is that you can lift the whole engine easily, as well as any of its parts. Don't try that on the family jalopy.

COMPRESSION PROBLEMS

When the engine misbehaves and tuning doesn't straighten it out, the first thing to do is test compression. This you do by turning the flywheel sharply counterclockwise, on the compression stroke. Disconnect the spark plug and tie the cable away from its post. Grasp the mower blade and jerk it backward. Do it twice to make sure you get the compression stroke— the exhaust stroke response will be weak. If the compression stroke doesn't jump back quickly, something is wrong. That can be a blown or leaking head gasket, valves not seating, or piston rings worn and passing oil. That condition also shows up as excessive oil use.

You must now remove the cylinder head. Remove the spark plug and any bracket or shroud that interferes with cylinder head removal. It won't be much. When you remove the head bolts, especially on a Briggs & Stratton engine, note that some are longer than others. It is vital to mark these bolts and their cavities; a long bolt in a short hole will break a fin, while a short bolt in a long hole will warp the head. Use paint to mark the bolts and holes.

When the cylinder head is off, clean out the carbon, scraping it and blowing it off. Examine the gasket carefully. If it is cracked or torn, and has telltale gas streaks across its surface, it means that you need a new gasket. Examine the valves, opening each one and looking carefully at the valve edges and their seats. Any sign of burning or wearing means a valve job. Look also at the piston surface and the rim and walls, moving the piston down to look at the walls. If there are heavy carbon deposits all around the combustion chamber, you may need new rings. If there is a pronounced ridge at the top of the piston cavity, that could be a sign of oil burning. It could also be benign. Turn the engine slowly and carefully to make sure that nothing is loose; turn it this way and that—back and forth. There should be no looseness whatsoever; everything should be tight and taut. Looseness means worn bearings and crankshaft, possibly piston rod.

The odds are heavy that the only thing wrong is carbon accumulation and worn head gasket. Once you take care of those items, replace the gasket and head, using *no* gasket cement. Head bolt tightening sequence is important and must be followed in Briggs & Stratton engines as Figures 145 and 146 show. Other makes have the same requirements; if you can't get a chart that spells out bolt tightening sequence exactly, just tighten opposite bolts, first tightening them lightly, then securely. In aluminum engines notice that you do not tighten the bolts much; otherwise, you could ruin the frame. Aluminum head bolts get tightened 140 to 165 inch-pounds; cast-iron torque is slightly higher. You can judge tightness by the feel of the bolts when you loosen them, and final tightening, without a torque wrench, should

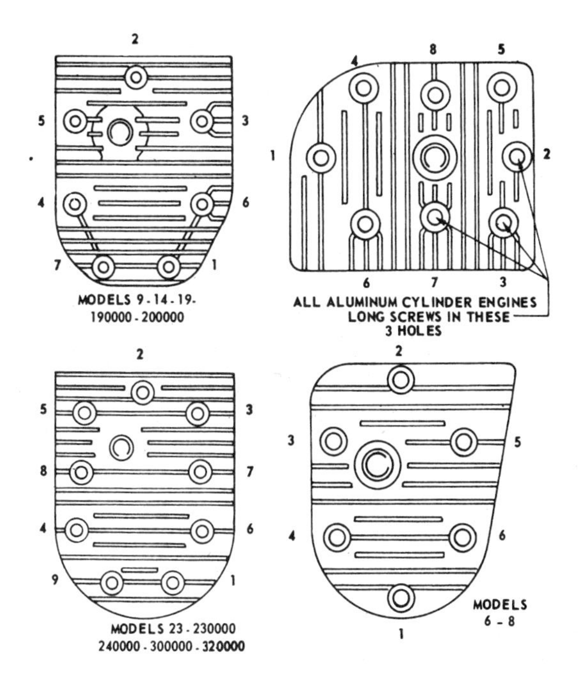

BASIC MODEL SERIES	IN. LBS. TORQUE
ALUMINUM CYLINDER	
6B, 60000, 8B, 80000 82000, 92000 100000, 130000	140
140000, 170000, 190000	165
CAST IRON CYLINDER	
5, 6, N, 8, 9	140
14	165
19, 190000, 200000, 23, 230000, 240000, 300000, 320000	190

FIGURE 145. Briggs & Stratton engine heads follow these patterns of head bolt removal and assembly.

be gauged according to the tightness you felt when you took them out. You don't need a torque wrench, unless you plan to go into the business.

If major engine work is indicated, you may not consider it worth the time and effort. You can buy new engines or good used engines at fairly low cost. However, if you want to learn to overhaul a small engine on the assumption that it will teach you to do the same job on an automobile, there are problems. The jobs are *different;* cars are less

TORQUE IN NUMERICAL ORDER

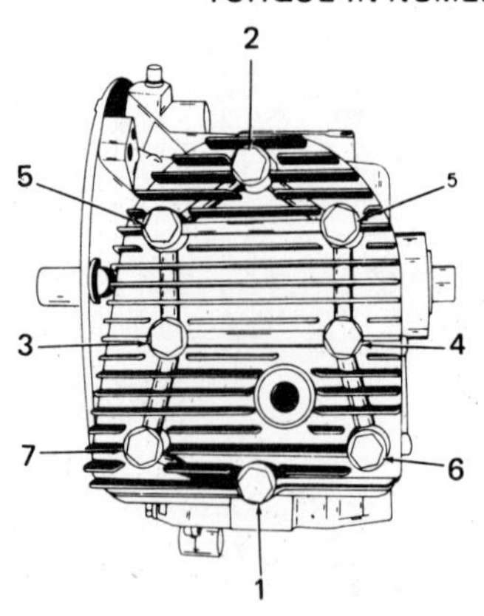

TORQUE TO 200 INCH LBS. IN 50 INCH LB. INCREMENTS

TORQUE IN NUMERICAL ORDER

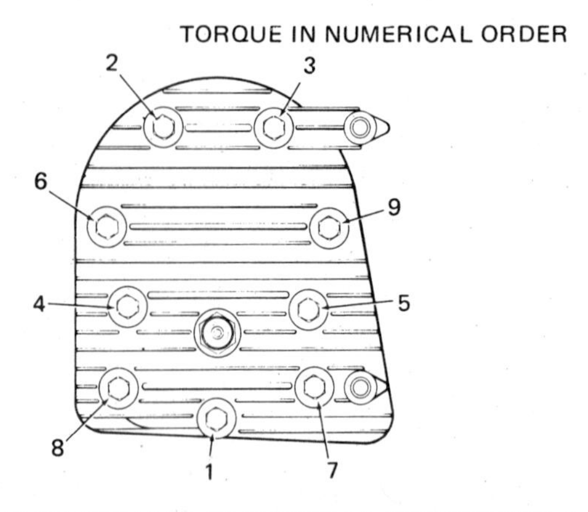

TORQUE TO 200 INCH LBS. IN 50 INCH LB. INCREMENTS

FIGURE 146. Tecumseh engine patterns of head bolt removal and tightening.

forgiving and much harder to tackle because of the number of parts involved and the amount of disassembly. Also, the economics differ. Overhauling an automobile engine is a big undertaking, for which you can pay up to $1,500, whereas when you deal with small engines you are dealing with much smaller numbers and tasks. Even though it may not be worth the trouble in terms of the task and the economics, if you want to learn to do the job, or like doing the job, then the economics are irrelevant.

You have examined the combustion chamber superficially, noting carbon formation and scraping it off. That's an easy, useful, sometimes restorative task. If you are interested in further engine disassembly, it is important to look more critically at combustion chamber carbon formation. If there is heavy formation around the intake valve, that's a certain indication of oil seepage through the valve lifter. If there are similar heavy, burned areas of carbon formation around the piston, and there is a burned ridge formation around the top of the piston wall, that indicates defective piston rings.

Don't just rush out to buy new rings and new valves. Engine rebuilding is an art and a science, and if it isn't terribly complicated, it does take some doing and some skill. If you've done any or all of the work discussed so far, you already have part of the skill.

You are now looking for the valves. Take the engine off the lawn mower deck, or wherever it normally functions. Remove the blade and drain the oil. To get that far requires unbolting any control brackets as well as the four (usually) bolts that hold the engine to the equipment, to say nothing of any plastic shrouds that may fit over the engine or part of it. Working on valves doesn't require taking all the ignition and fuel systems off, but you will soon discover that it's much easier to work on the engine underneath if you take the top parts off. So take them off before you get serious about the valves. That means taking off the starter, flywheel, etc., as before.

Put the engine, now drained of oil, somewhere that offers good light and convenience. Unbolt the bottom oil pan (or sump, as it is called), and beware of the messy consequences—everything has oil on it, even though you've already drained the engine. Keep a lot of rags around and do the unbolting of the sump on newspapers.

To remove valves requires a valve lifter—actually a compressor that pushes the spring together so you can take out the valve keepers, which are collars or a pin and cups. A screwdriver may work instead of the valve compressor, but if it doesn't the compressor is an inexpensive tool. It fits over the top and bottom of the spring. If you use it, adjust the jaws so that they just touch the top and bottom of the valve chamber.

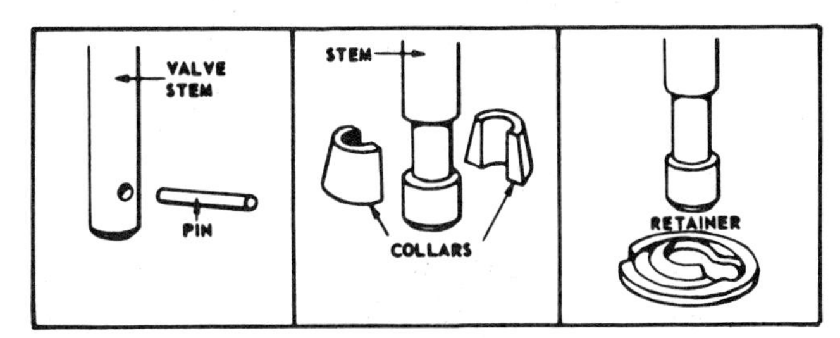

FIGURE 147. Three methods used by Briggs & Stratton to hold the valve spring retainers.

That will keep the upper jaw from getting into the coils of the spring and messing it up. Tighten the compressor sufficiently to allow you to pull out the collars or pin and lift the valve out. Then lift out the valve compressor and spring, and release the spring. Keep all the parts together.

If you deal with valves using retainers instead of collars or pins, slip the upper jaw of the spring compressor over the top of the valve chamber, and the lower jaw between the spring and retainer. Compress the spring and remove the retainer. Pull out the valve, compressor and spring. In this case the screwdriver will work just as well; in the other cases it won't.

In some engines the exhaust and intake springs are identical; in others they aren't. So keep them apart.

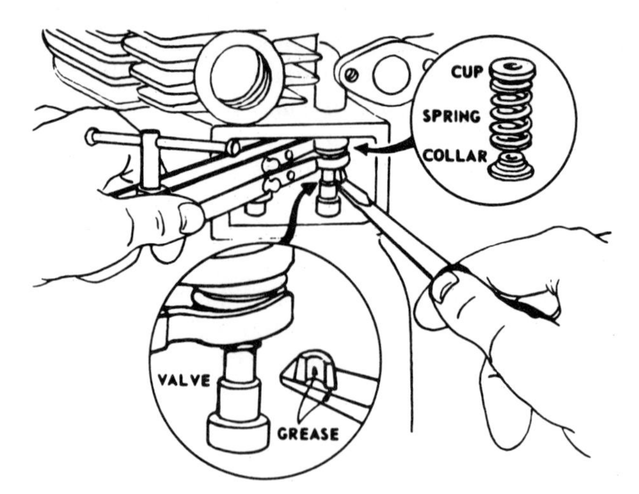

FIGURE 148. Valve spring removal using lifter.

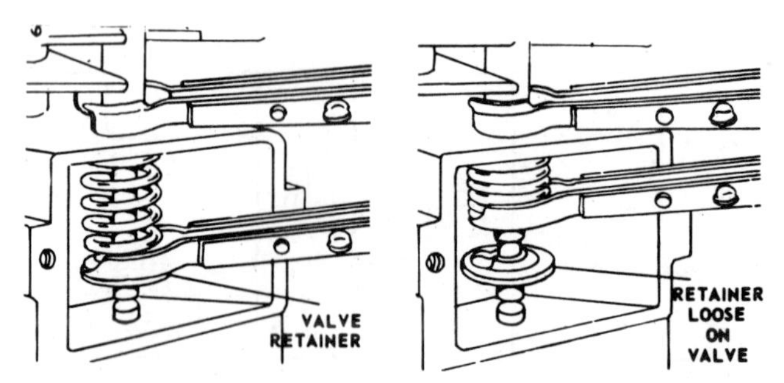

FIGURE 149. With compressor in place, the valve retainer can be pulled out.

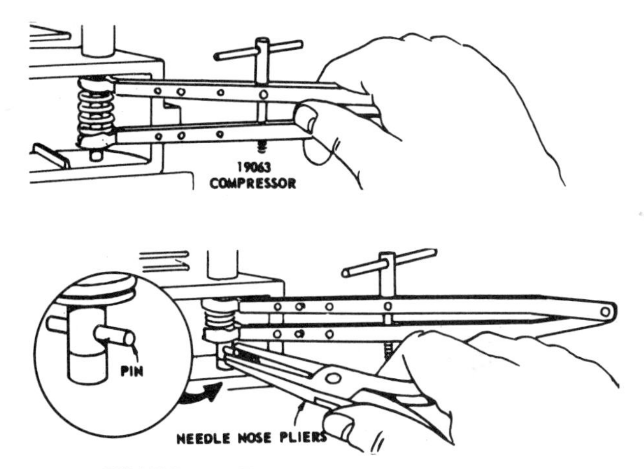

FIGURE 150. The retainer out, pull out the valve, then lift out the spring.

Professionals approach valve jobs with a lot of tools and precise measurements in mind, but you can see most of what is wrong with valves. Put the valve back in its hole and try to determine how close the fit is. It should be tight. If you can detect any looseness at either end, or in any part of the valve travel, the valve guide will have to be replaced since guides are made of softer metal than are valves. Replacing valve guides is not a job you can do.

You need to measure the valve surfaces and valve seat. If they are worn or burned, they will need either grinding or replacing. Valve seat margins (top of seat) should be between $^{3}/_{64}$ inch and $^{1}/_{16}$ inch (1.2 mm to 1.6 mm). Valve margin (top of valve) should be $^{1}/_{32}$ inch; discard if $^{1}/_{64}$ inch or approaching it.

Valves and seats face at a 45° angle. If you are reusing the old ones, they will need resurfacing at a machine shop. After resurfacing, or if you are replacing, you need to do something called lapping the surfaces. This consists of smoothing the surfaces beyond what you can do with sandpaper. That process requires that you buy some lapping compound and smear it on the opposing surfaces of the valve and valve seat. Then, turn

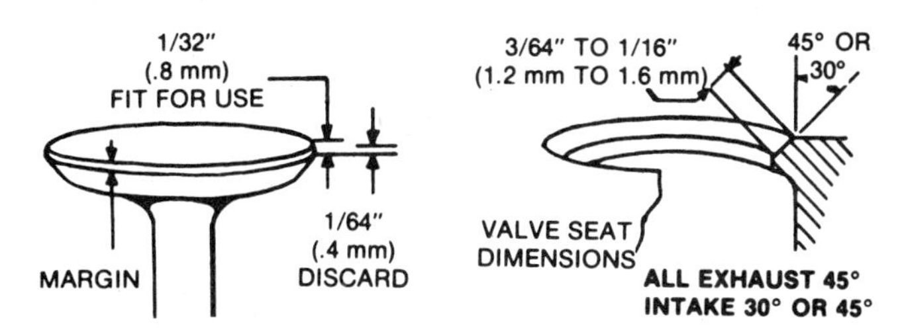

FIGURE 151. Valve and valve seat margins.

the valve around by hand, either by using a lapping tool (which you can rent or buy—they don't cost much) or by improvising a tool. Take a small stick and put a suction cup at the end of it that is capable of holding the valve firmly enough to turn it against the valve seat. Turn the valve around until you get a polished surface. You can start with a coarse lapping compound if you see imperfections in one or both surfaces. Finish with a fine compound. Don't lap too much; too many laps spoil the combustion. Valves have to seat in cold and hot engines, so lapping too much can throw off the relationship between valve and seat, which changes with engine temperature.

If you are buying a new valve, the chances are good that you need a new seat. First do the measuring as indicated above. If the valve seat isn't worn down too much, it may need nothing more than refacing, which lapping could correct. If you need a new seat, the old one must be gotten out. Valve seat pullers are available, naturally. They grab the seat from below and pull it up. They're Simple Simon tools, but they're hard to improvise, though that, too, can be done.

Lapping and refacing are two different treatments. Refacing can be done only on a lathe, which cuts off a skin-thin layer of metal. You still have to lap the two surfaces together. Slipshod methods won't do. If the valve and seat fail to match perfectly, they will burn up quickly. So, it is important to measure valve and seat surfaces and take appropriate action—new if worn below numbers given above; resurfacing or refacing, if imperfections are marked, followed by lapping. New valves and seats also require lapping.

The valve seat puller is also a valve seat installer in some cases—it forces the new seat in. However, if you use the right tactics to absorb force, you can pound the new seat in. A block of wood, a lead hammer, or

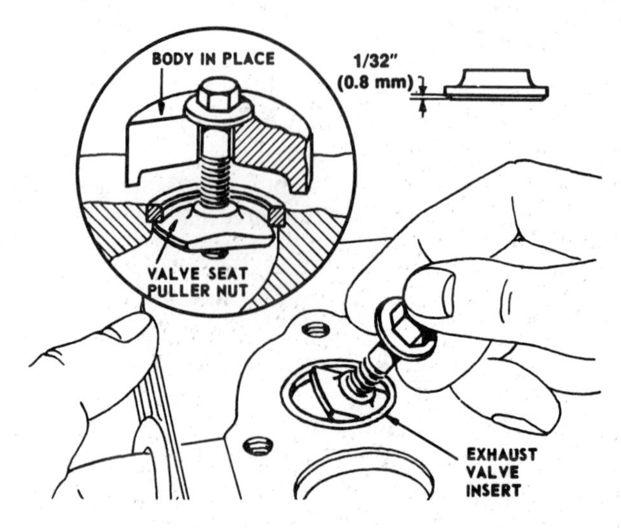

FIGURE 152. Valve seat puller insertion.

a heavy rubber hammer can be used to start the seat into its hole. Then use something that covers the seat over its diameter to pound it in completely. Wood is preferable to avoid damaging the seat.

If the valve guide is worn, you will have to take the engine to a machine shop for the installation of a new bushing and guide, but put the new valve in before you do anything else, and test it for tightness. If it isn't loose, the chances are good that you are home free.

A new valve or valves will require that the stem or stems be ground off to obtain the correct clearance between the valve stem and the tappet that pushes it. The tappets ride on camshaft cams. Turn the crankshaft to top dead center—just past the compression stroke—piston at top, both valves fully closed. To turn the crankshaft you need to turn the flywheel, or turn from below. You have removed the flywheel, but put the clutch back on (or put a nut on the crankshaft stem where the flywheel anchors) and turn that—or simply set the flywheel back on temporarily.

Briggs & Stratton clearances on their engines are given in Figure 153. Any differences must be corrected by grinding off the new valve stem (they will always be too long if there are differences). If you are using old valves and seats, the probability is that you will have nothing to do to the stems.

When grinding off the stems, it must be done straight across. Put the valve in a vise with the amount of stem to be ground off sticking above the vise jaws. File it off and smooth the surface. Note that both valve and valve seats control the valve tappet clearance, so any metal taken off either or both surfaces will change tappet clearance.

When replacing valves, reverse the disassembly procedure. If the springs differ, the heavier one goes on the exhaust valve. Examine the springs carefully before replacing them. Buy new ones if the springs have obvious damage—cracks, broken pieces, etc.

When the valve retainers are held by a pin or collars, put the spring, retainer, and cup into the valve spring compressor. Compress the spring until it is solid. Put the compressed spring, retainer, and cup (when that is used) into the valve chamber. Drop the valve into place, pushing the stem through its retainer. Hold the spring up in the chamber, with the valve down. Insert the retainer pin, using a needle-nose pliers, or place the collars in their groove in the valve stem. Lower the spring until the retainer fits around either pin or collars, then pull out the spring compressor.

If a self-locking retainer is used, compress the retainer and spring. The large diameter of the retainer should face toward the front of the valve chamber. Put the compressed spring and retainer into the valve chamber. Drop the valve stem through the larger area of the retainer slot. Move the compressor to center the small area of the valve retainer slot onto the valve stem shoulder. Release spring tension and remove the compressor.

MODEL SERIES	INTAKE				EXHAUST			
ALUMINUM CYLINDER	MAX.		MIN.		MAX.		MIN.	
	Inches	Milli-meter	Inches	Milli-meter	Inches	Milli-meter	Inches	Milli-meter
6B, 60000, 8B, 80000, 82000, 92000, 94000, 100000, 110000, 130000, 140000, 170000, 190000, 220000, 250000	.007	0.18	.005	0.13	.011	0.28	.009	0.23
CAST IRON CYLINDER	Inches	Milli-meter	Inches	Milli-meter	Inches	Milli-meter	Inches	Milli-meter
5, 6, 8, N, 9, 14, 19, 190000, 200000	.009	0.23	.007	0.18	.016	0.41	.014	0.36
23, 230000, 240000, 300000, 320000	.009	0.23	.007	0.18	.019	0.48	.017	0.43

FIGURE 153. Valve tappet clearance table, by Briggs & Stratton.

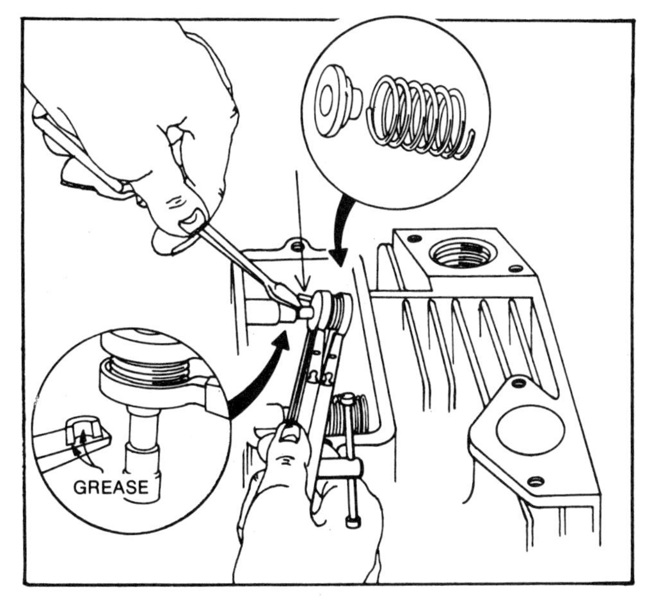

FIGURE 154. Compressing valve spring for installation. This is an exhaust valve.

In normal use, and with the equipment that doesn't have to do heavy-duty, high-speed work over long periods of time, valves will last for years. However, standard exhaust valves often burn up when a piece of combustion deposit lodges between the seat and the valve face, preventing the valve from closing completely. Two cures are possible: a Rotocap or valve rotator, which turns the exhaust valve slightly on each lift, wiping off any deposits, or a Stellite exhaust valve, which has greater resistance to heat. Any dealer can sell you such devices; now you know how to install them. You buy them by engine model number.

TECUMSEH VALVES

In addition to the valve configuration discussed, Tecumseh and others make valve-in-head small engines. These are miniature automobile engines. The valves need periodic adjustments. Adjustment is at the rocker arms, as in cars.

Remove the top (it's called a breather) and then the valve cover. The push rods must be at their lowest position; the piston at top dead center with both valves completely closed and both push rods at their lowest position, while the rocker arms (the other ends) will be at their highest position.

With the engine cold, use a feeler gauge between the valve stem and the rocker arm of both intake and exhaust valves. Exhaust valve clearance is .010 inch; intake valve is .005 inch. Open the locknuts and turn the adjusting screws until the clearance is correct. Then tighten the locknuts. Turn the adjusting screw clockwise to decrease clearance, counterclockwise to increase it.

If valve grinding and other major service is required on Tecumseh or any other valve-in-head engine, it is rather more complicated than what we've been doing thus far. It's an auto engine overhaul in miniature. In some ways it is more difficult—valve guide service, for example. Nonetheless, you can remove the cylinder head for examination of the condition of the valves and seats, cleaning out carbon deposits in the process. If you need major valve surgery, such as that undertaken with the Briggs & Stratton engines, you had best take your engine to a shop. Such surgery involves too many complications, tools, and procedures.

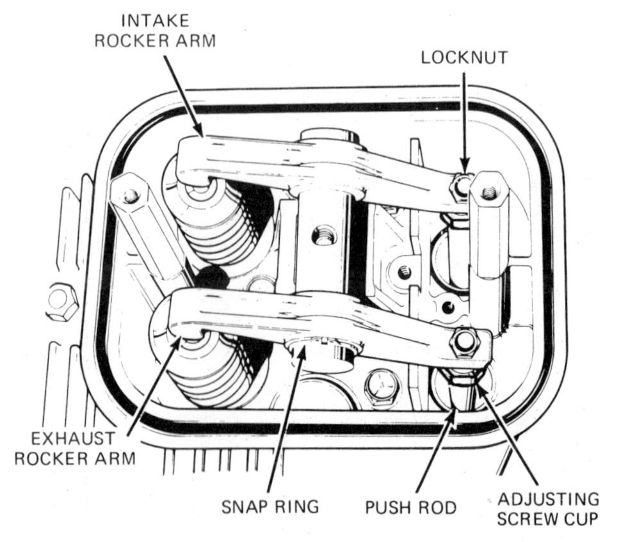

FIGURE 155. Setting up Tacumseh valves for lash adjustment.

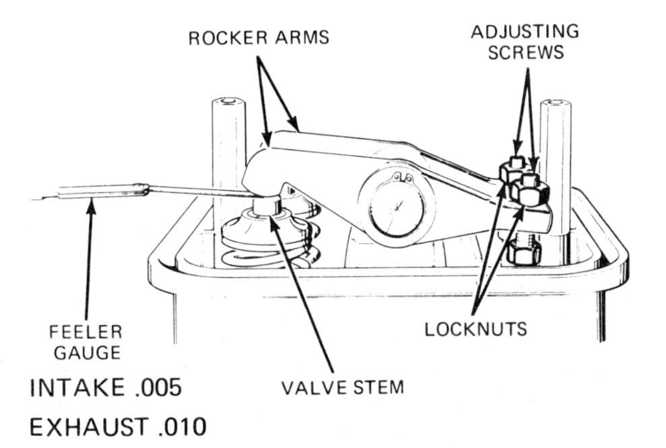

FIGURE 156. Measuring and adjusting valve lash on Tecumseh four-cycle engines.

Tecumseh two-cycle engines have reed valves, which are readily serviced. Mostly what is required is cleaning. Dirt and other deposits prevent the reed valves from sealing, thus drastically interfering with engine efficiency. Reeds should not bend away from their plates by more than .010 inch. Usually, when they do, they must be replaced, but often dirt is the culprit and can be scraped off, sanded off, or removed with solvent. Reed valves are also found on air compressors, at the cylinders. They, too, need cleaning periodically.

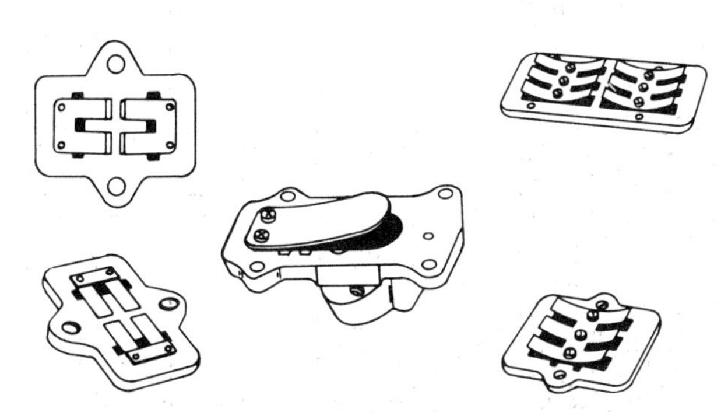

FIGURE 157. Reed valves on Tecumseh, Clinton, and other engines, power snowblowers and many other power tools. The valves should be cleaned periodically.

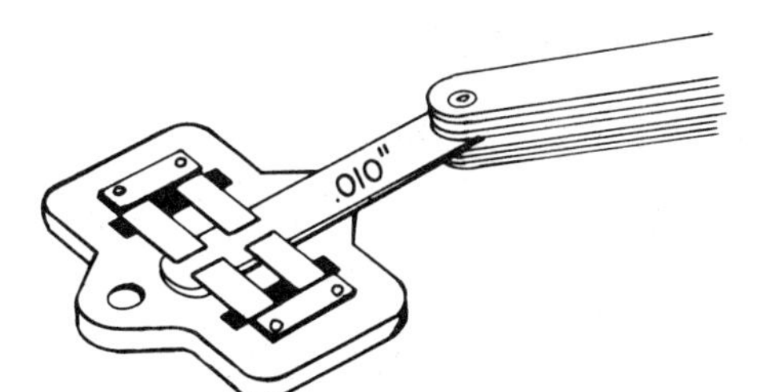

FIGURE 158. Use a feeler gauge to check extent of reed valve travel away from their plates.

PISTON SERVICE

Piston ring wear is indicated primarily by oil burning and poor compression. We have called attention to such visual inspections as the carbon formations on top of the piston and at the top of the cylinder wall (the ridge around the top). If there is heavy carbon formation on top of the piston and at the ridge and around the combustion chamber, and the valve guides are exonerated, the piston rings are usually at fault. If the engine fails the test of pulling the crankshaft against the compression stroke—the engine fails to snap back smartly—that is further and conclusive evidence of weak compression. The fault can be in the valves and head gasket; we've addressed those areas. Now on to the piston and its problems.

Generally, rings outwear valves, so it would be the unusual engine whose poor compression would fail to respond to the work thus far. However, piston rings and bearings are major wearing components.

In Briggs & Stratton engines and other similar engines, you begin disassembly of the piston by bending down the connecting rod lock—a metal tab you push down with a punch or screwdriver—and then hammer on each end of the connecting rod cap. Next, unbolt the cap and the piston is free to come up and out—if the ridge at the top of the cylinder wall is reamed. A special tool can be rented to do that but you can also use a small hand drill and stone to do the job. Once the piston and piston connecting rod are out, remove the piston pin lock with a needle-nose pliers (one end of the pin is drilled to permit removal). Remove the rings one at a time, marking their position on the piston and the direction they face. Again, there is a ring expander designed to prevent damage to rings as you remove them. How do you remove them? *Very* carefully; they are delicate. That's if you're not going to install new rings. It isn't automatic. You examine the rings and test them.

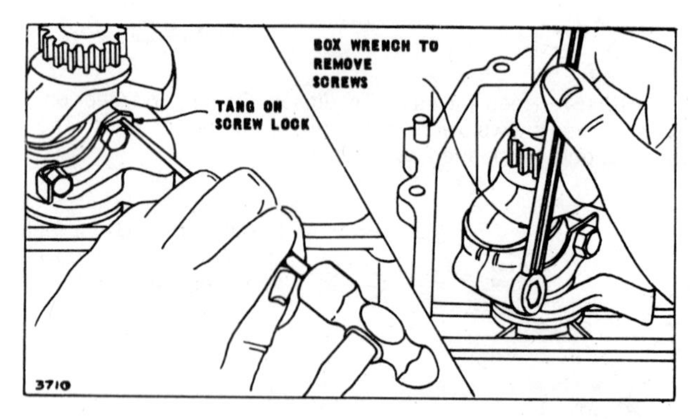

FIGURE 159. Bending open the lock on the connecting rod.

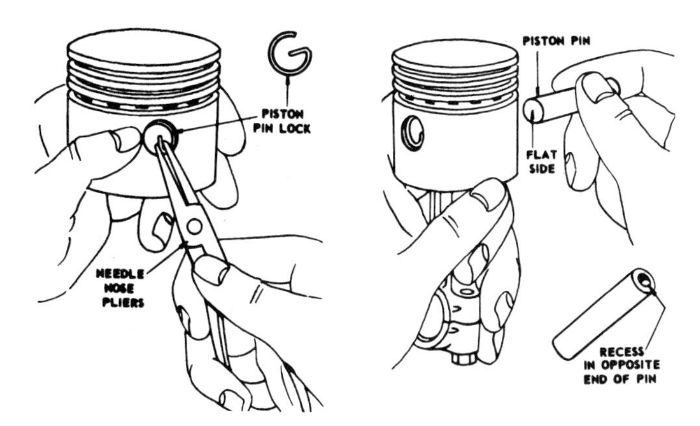

FIGURE 160. Removing piston lock and pin.

First, examine and test the cylinder walls and piston. The walls should be free of scoring and other rough wear. The piston must first be cleaned of carbon from the top ring groove. If you have an old ring, you can use that to clean out the carbon, and clean out the oil holes, too. Then, place a new ring on the groove, and check the remaining space in the groove with a feeler gauge. If you can put a .007-inch feeler gauge between the new ring and the groove, the piston should be replaced.

To check rings, clean all carbon from the rings and the cylinder bore. Push each old ring, one at a time, an inch down into the cylinder. Check the gap between the two ends. If the gap is greater than that shown in the table in Figure 163, you need new rings. Note that there are different specifications for aluminum and cast-iron cylinders.

In aluminum cylinder engines you do not deglaze cylinder walls when installing rings. When you use chrome rings, either in aluminum or cast-iron models, no honing or deglazing is required. Also, cylinder bores can be a maximum of .005 inch oversize when you use chrome rings. It's something to think about. Honing and deglazing cost money.

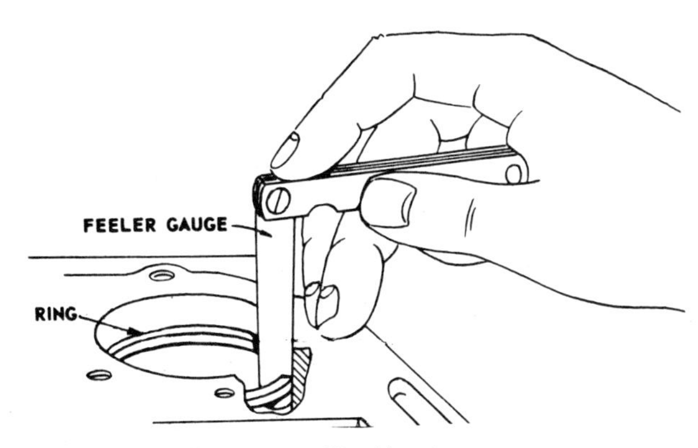

FIGURE 161. Checking ring gap.

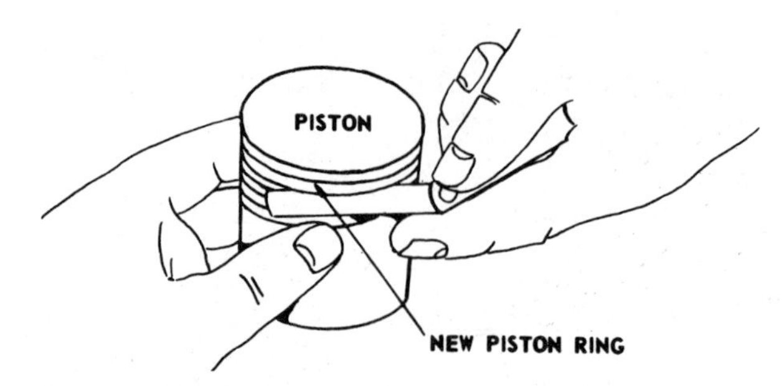

FIGURE 162. Checking ring grooves for wear.

<table>
<tr><th>BASIC MODEL SERIES</th><th colspan="2">COMP. RING</th><th colspan="2">OIL RING</th></tr>
<tr><th>ALUMINUM CYLINDER</th><th>Inches</th><th>Milli-meter</th><th>Inches</th><th>Milli-meter</th></tr>
<tr><td>6B, 60000, 8B, 80000</td><td rowspan="4">.035</td><td rowspan="4">0.80</td><td rowspan="4">.045</td><td rowspan="4">1.14</td></tr>
<tr><td>82000, 92000, 110000, 111000</td></tr>
<tr><td>100000, 130000</td></tr>
<tr><td>140000, 170000, 190000, 250000</td></tr>
<tr><th>CAST IRON CYLINDER</th><th>Inches</th><th>Milli-meter</th><th>Inches</th><th>Milli-meter</th></tr>
<tr><td>5, 6, 8, N, 9</td><td rowspan="5">.030</td><td rowspan="5">0.75</td><td rowspan="5">.035</td><td rowspan="5">0.90</td></tr>
<tr><td>14, 19, 190000</td></tr>
<tr><td>200000, 23</td></tr>
<tr><td>230000, 240000</td></tr>
<tr><td>300000, 320000</td></tr>
</table>

FIGURE 163. Ring gap rejection table on Briggs & Stratton engines (others are similar).

The piston rod has a crankpin bearing, which, if it is worn or scored, means a new piston rod. The crankpin bearing hole and pin bearing hole must meet "rejection" specifications, shown in a table in Figure 164. If there is wear in the piston pin bearing, you can get oversize pins—.005 inch oversize. Remember, the crankpin bearing in the rod, if worn, means a new connecting rod. Also check the piston pin and see how it stacks up against the rejection size specifications.

Now examine the cylinder walls carefully. Cylinders with wear up to .005 inch can be O.K. with new chrome ring sets, and as indicated, they require no honing or glazing. So get chrome, if all else is well, cylindrically speaking. To determine if all else is well, you need an inside micrometer or telescoping gauge.

If the cylinder bore is worn beyond the .005-inch limit, either buy a new engine or find out what a shop will charge to rebore the cylinder. It is not something you can do; it requires a lot of equipment of a

BASIC MODEL SERIES	PISTON PIN		PIN BORE	
ALUMINUM CYLINDER	Inches	Milli-meter	Inches	Milli-meter
6B, 60000	.489	12.42	.491	12.47
8B, 80000	.489	12.42	.491	12.47
82000, 92000, 110000	.489	12.42	.491	12.47
100000	.552	14.02	.554	14.07
130000	.489	12.42	.491	12.47
140000, 170000, 190000	.671	17.04	.673	17.09
220000, 250000	.799	20.29	.801	20.35
CAST IRON CYLINDER	Inches	Milli-meter	Inches	Milli-meter
5, 6, 8, N	.489	12.42	.491	12.47
9	.561	14.25	.563	14.30
14, 19, 190000	.671	17.04	.673	17.09
200000	.671	17.04	.673	17.09
23, 230000	.734	8.64	.736	18.69
240000	.671	17.04	.673	17.09
300000, 320000	.799	20.29	.801	20.35

FIGURE 164. Piston pin rejection sizes.

specialized sort, including a drill press. It isn't likely that your lawn mower or snowblower will need radical surgery on its cylinder, but if it is burning oil and needs new rings, the probabilities are high that new chrome rings will fill the bill.

Before you reassemble the piston rod and piston, you have to make certain that the ridge at the top of the cylinder is gone; otherwise, you will ruin the new rings at installation time. So use a drill and grinding stone if emery cloth or sandpaper won't do it.

The approved way to remove old rings and install new ones is to use the special tools. If you're going to do it by hand, use a small screwdriver, pull one end of the ring up, and slowly pull the entire ring out of its groove, carefully avoiding scratching the piston. Then, pull the ends slightly apart and work the ring off, either up or down. The upper rings come off the top; the bottom rings should be worked off the lower or skirt end of the piston. If you can get them all off this way, without scratching the piston and without breaking the rings, you can get the new ones on the same way. You're trying to make an omelet without breaking the eggs, but if you break the old ones, don't risk the new—rent or buy a ring compressor. They're not terribly expensive. The fragility of these rings is amazing when you consider the wear they get year after year.

The piston must be cleaned out thoroughly, especially the oil holes in the grooves. You can use a small drill bit on the oil holes; an old

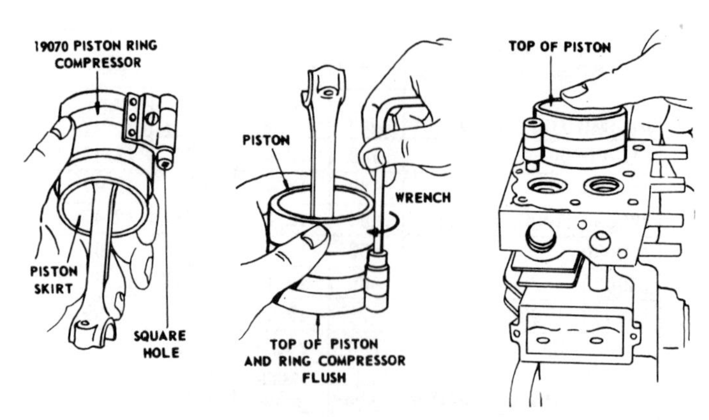

FIGURE 165. Clinton piston ring positioning.

ring makes the perfect carbon scraper for the grooves. A screwdriver blade of the right size will work if you don't have an old ring, but use it cautiously.

To put it all together, after testing the cylinder bore, push the piston pin into its position in piston and rod. Place the pin lock in the groove at one side of the piston. From the opposite side of the piston, insert the pin—flat end first with the solid pin, either end when a hollow pin is used—until it stops against the pin lock. Use a needle-nose pliers to assemble the pin lock in the recessed end of the piston. Locks must be firmly set in grooves.

Each ring has its own function and place and correct installation. If they are wrong, they won't work. The instructions that accompany new

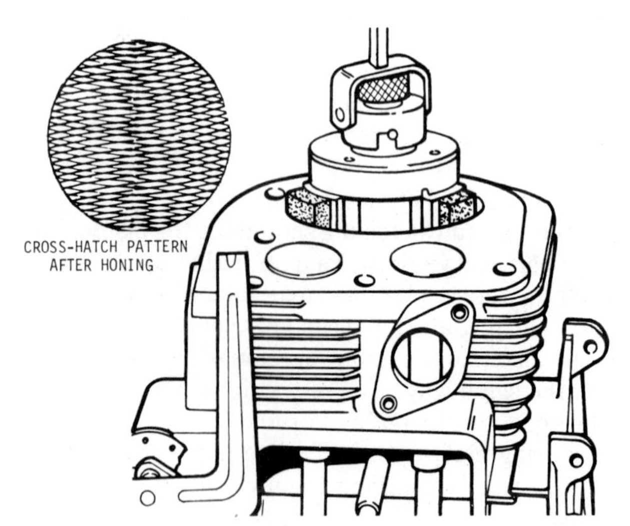

FIGURE 166. An out-of-round check of cylinder walls will save you a lot of grief. If they are defective, honing and cross-hatching will renew them.

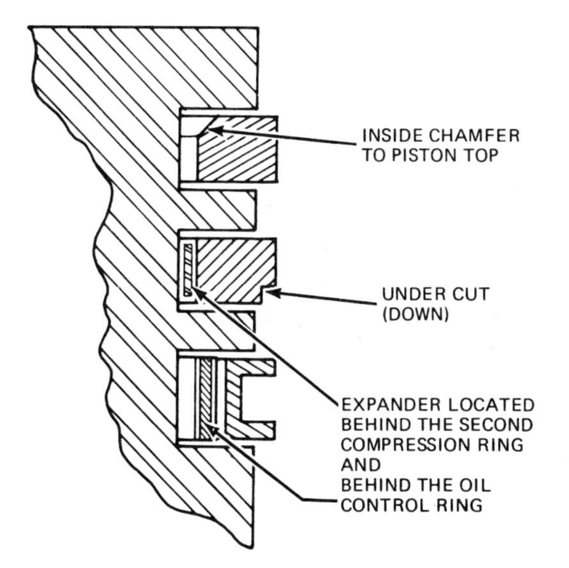

FIGURE 167. Tecumseh ring installation profile.

rings will usually clue you in, but don't count on that. First, make a diagram of the old rings when you take them out, marking the edge that goes up, the edge that goes down, the configuration of the ring that is first, second, and third.

Look at the Tecumseh ring installation. Here you can see clearly the rather small but crucial differences between rings. Tecumseh engines require that the cylinder be reconditioned and deglazed. Tecumseh says you can deglaze the cylinder wall with a fine abrasive cloth, thus helping to seat the new rings. Note marks on the first and second rings that indicate the top of the ring. Ring end gaps must be staggered;

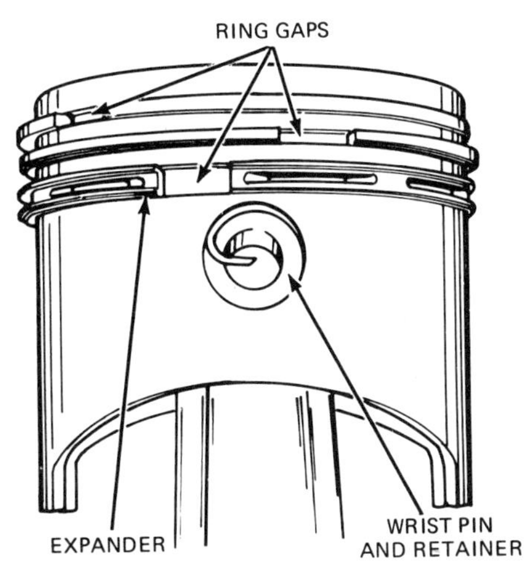

FIGURE 168. Tecumseh ring gap relationships inside the ring grooves.

otherwise, you are building in compression loss. Also, use plenty of engine oil to lubricate all friction surfaces in the assembly process. This is true of all engines.

In the Tecumseh connecting rod installation, if you are putting the old one back in, use new nuts on the bolts. These are locking nuts.

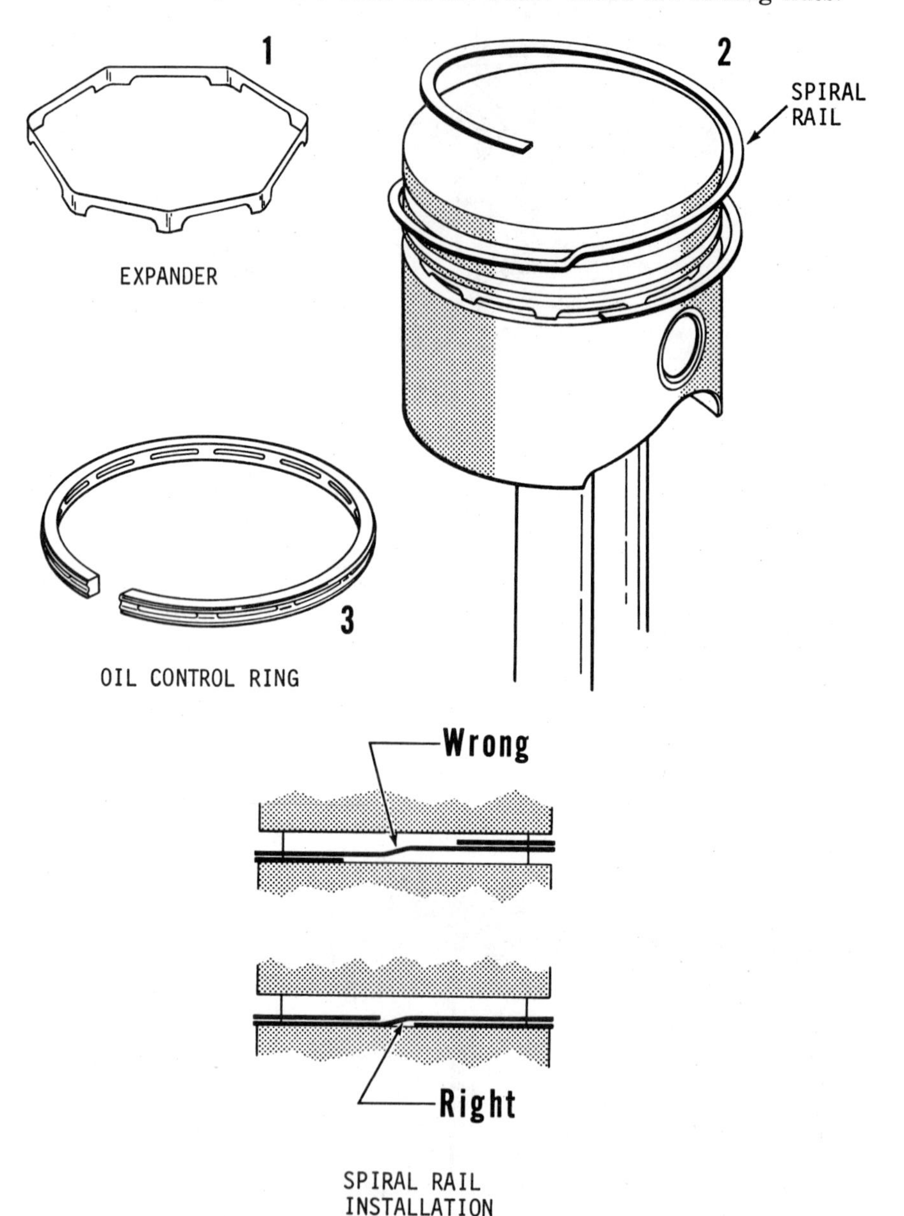

FIGURE 169. Kohler uses a spiral rail with the oil control ring (top), along with a ring expander. It is installed as illustrated.

Tecumseh stresses the importance of new ones, to retain torque. Install the connecting rod with the match mark facing out of the cylinder toward the power takeoff end of the crankshaft (where the blade or whatever is attached). Through-bolt heads on the connecting rod must be seated tight against the shoulder on the rod. This, too, is important. Torque is 110 inch-pounds, if you have a torque wrench. If not, tighten securely. Then hit the rod bearing cap above each locknut with a drift (punch) and hammer, to seat the cap. Give the nut another tightening; a torque wrench would reveal that the strike, though it should be light, had loosened it. Don't overtighten.

Clinton rings call for the oil ring at the bottom, which is normal, with an expander that goes on first (also common but not invariable), and next the "scraper" ring in the middle, which has a step machined out of the lower outer circle and faces down to the skirt end. At the top is the compression ring. It, too, has its own configuration.

Different engine manufacturers have different details of valve and piston fastening and components. Because these are among the keys to engine performance, it is not strange that everyone tries to come up with a new detail that will make a magic step forward—in performance, durability, simplicity, or whatnot. It doesn't happen, but that doesn't keep people from trying. Someday it might.

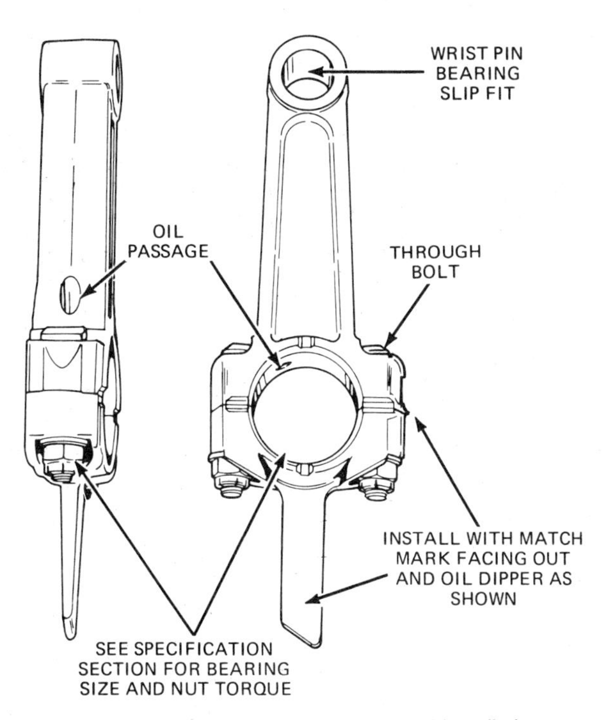

FIGURE 170. Tecumseh connecting rod installation.

CRANKSHAFTS AND CAM GEARS

Crankshaft wear occurs slowly; signs are rarely dramatic. Engine vibration that can't be traced to the blade or other attachments, or simply loose mounting bolts or incorrect mounting of the engine, may finger the crankshaft as the culprit.

Crankshaft removal differs from make to make and model to model. Consider Kohler engines with the so-called dynamic balance system. These engines—K241 and K301 models, and on all K321 and K341 models—have two balance gears that run on needle bearings. They turn on two small shafts that are pressed into special "bosses" in the crankcase. Snap rings hold the gears; they are driven off the crank gear in the opposite direction to the crankshaft. When you go trouble-shooting these engines for vibration, you have to check out these gears before you tackle the crankshaft itself. You have to check them all out.

If one of the small "stub" shafts that holds the two balance gears becomes worn, it can cause the opposite of dynamic balance, which we may conveniently label as a wobble—a nontechnical term. You have to replace the stub shaft by pressing it out and pressing in a new one. These stub shafts vary in height. One of them protrudes about 11/16 inch above the shaft base; the other about 1 1/16 inch. On this one you must use a 3/8-inch spacer next to the base (it's also called a boss).

If you have to remove these shafts and gears, and the crankshaft, you have to time and align the balance gears when you complete the job. There is a special, inexpensive tool that aids this process, but without the tool it goes this way:

Put the crankshaft back into the block. Align the primary timing mark on the top balance gear with the standard timing mark next to the crank gear. Press the crankshaft until the crank gear is engaged 1/16 inch into the top gear (narrow side of gear). Turn the crankshaft to align timing marks on the crank gear and cam gear, then press the crankshaft the remainder of the way back into the block.

Now the bottom balance gear assembly must be aligned. Align the secondary timing mark on the bottom gear-bearing assembly with the secondary timing mark on the counterweight of the crankshaft. Install the gear-bearing on the shaft. The secondary timing mark will also be aligned with the standard timing mark on the crankshaft after its installation. Washers and spacers that came off in disassembly must go back on. They include one .010-inch spacer over the stub shaft before you install the bottom gear-bearing assembly, and one .005-inch and one .020-inch spacer—the larger next to the retainer—to obtain correct end play of .002 to .010 inch.

An easier way to do all this is to buy an inexpensive Kohler assembly tool, Y357. With that tool, follow these steps:

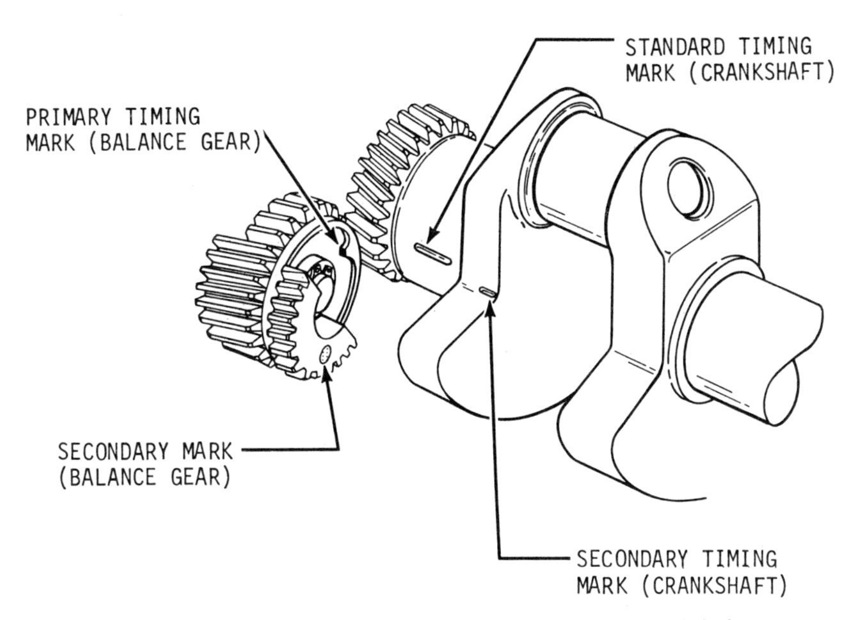

FIGURE 171. Timing marks on the balance gears and crankshaft must be aligned correctly.

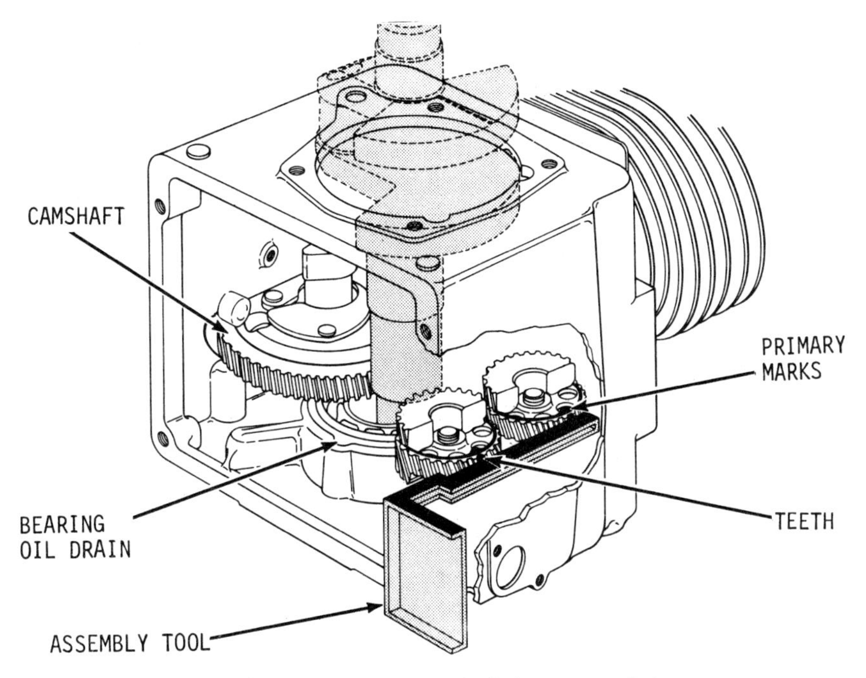

FIGURE 172. The easy way to time the balance gears is to use a simple timing assembly tool.

1. Turn both balance gears so that primary timing marks line up with teeth on tool, then insert tool in mesh with gears.
2. Hold gears with tool butted against gasket surface, align standard marks on crankshaft with bearing oil drain hole, then lower shaft until the crank gear is just started into mesh (about 1/16 inch) in balance gears.
3. Remove tool, align crankshaft-camshaft timing marks, then press crankshaft all the way into the crankcase.
4. As a final check, turn the crankshaft to see if standard timing mark on crankshaft lines up with the secondary timing mark on the bottom balance gear. If these marks cannot be lined up, timing is off and you must start over.

On Briggs & Stratton aluminum cylinder engines, crankshaft removal requires the filing or sanding of rust or burrs from the power takeoff end before removing the oil sump (crankcase cover). If the cover sticks, tap it with a soft hammer on opposite sides. Turn the crankshaft to align the crankshaft and cam gear timing marks. That's important. Lift out the cam gear, then remove the crankshaft. On models with ball bearings, the crankshaft and cam gear must be removed together.

Cast-iron cylinder engines have slightly different crankshaft removal procedures. These include model series 5, 6, 8, and N. First, plain bearings. Remove the magneto. Remove burrs and rust from the power takeoff end of the crankshaft. Remove cover and crankshaft. Model series 6FB, 6FBC, 6SFB, 8FB, 8FBC, and 8FBP, with ball bearings: Remove the magneto. Drive out the cam gear shaft while you hold the cam gear, to prevent dropping it. Push the cam gear into the recess, and pull crankshaft out from the magneto side. Other models are similar.

It goes without saying that you are working on the crankshaft after you have removed valves and piston. The crankshaft drives the piston rod. It won't come out until you unbolt the rod, but each model series differs slightly. A little study will help you puzzle it out.

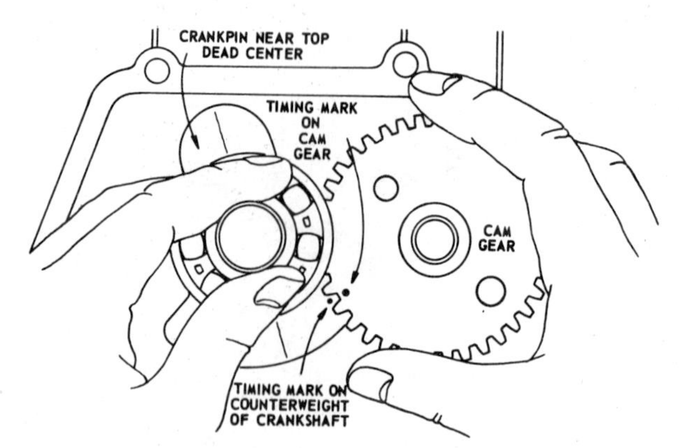

FIGURE 173. Engines by Briggs & Stratton with ball bearings have these timing mark relationships.

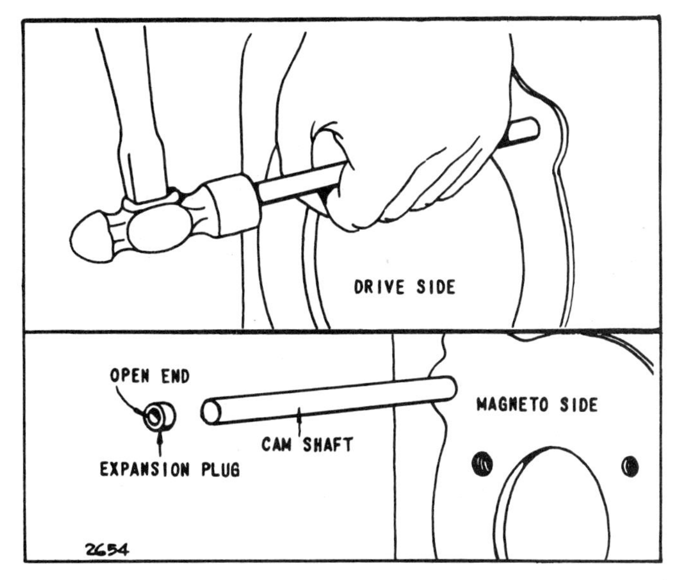

FIGURE 174. Camshaft removal on Briggs & Stratton engines.

Crankshaft wear may be obvious, with burrs, scratches, broken or badly scarred portions. These are enough to warrant a new unit, but wearing away of the key surfaces past certain specifications also disqualifies a crankshaft.

The cam gear must be inspected for wear and nicks on the gear teeth. But the cam gear also wears down, and there are rejection tables that disqualify cam gears.

One test you can make on cam gears is the automatic spark advance on models with "Magna-Matic." Place the cam gear in its normal

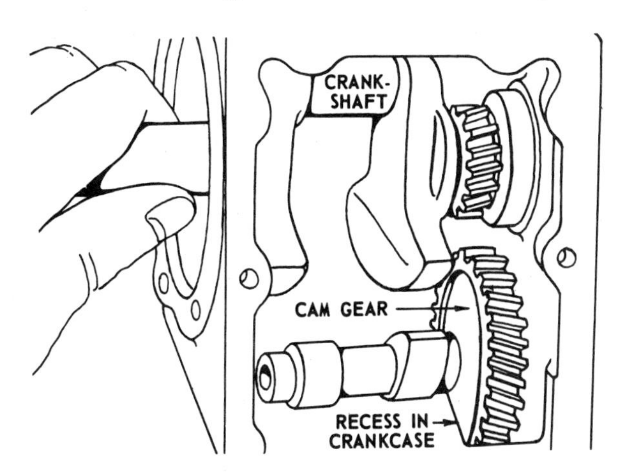

FIGURE 175. Crankshaft installation and removal on Briggs & Stratton engines.

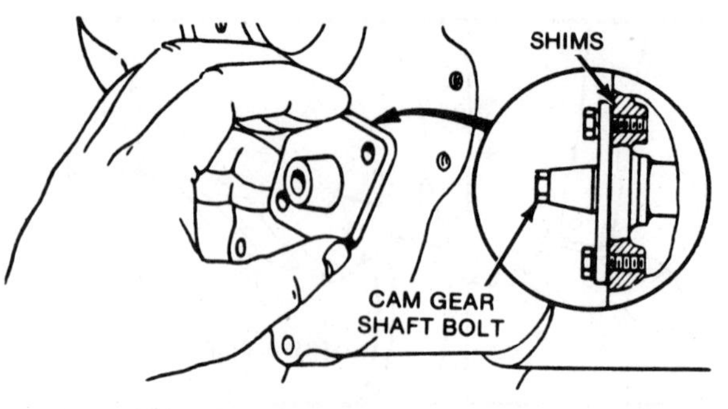

FIGURE 176. Cam gear removal.

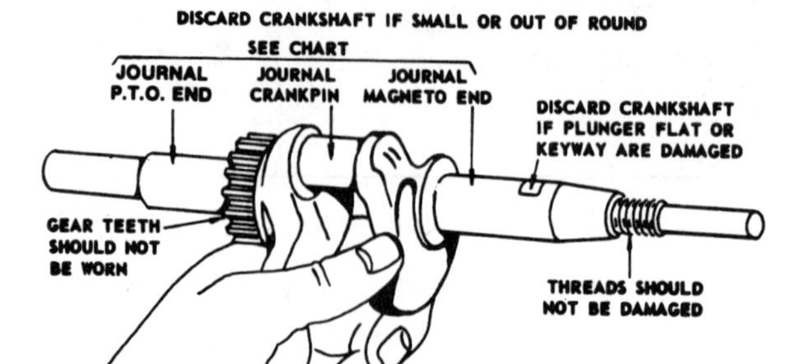

FIGURE 177. Visual checks on crankshaft wear.

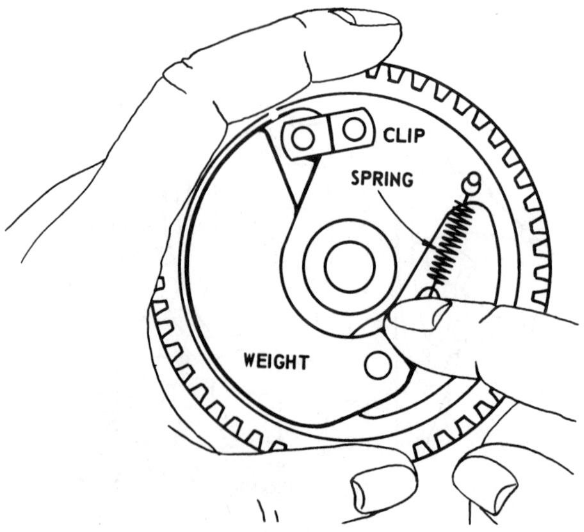

FIGURE 178. Automatic spark advance check on Briggs & Stratton engines with "Magna-Matic."

operating position with the movable weight down. Push the weight and release. The spring should lift the weight. If it does not, the spring is stretched or the weight is binding. The spring should be replaced if lubrication and cleaning of the weight doesn't cure the problem.

Ball bearings on crankshafts are special problems. If, when the crankshaft comes out and the bearing is faulty—a strong possibility when noticeable engine vibration has caused you to remove the crankshaft—the bearing must be pressed off. Only an arbor press will do the job, though you might try using a wood block and pounding on that. However, you can't use metal on the crankshaft to pound off the bearing (despite my recommendation to pound on the other end of it to get the flywheel off) without risking damage to it. That's because the press fit is tight for a long space and requires too much force. If you take the crankshaft to a shop for the job of pressing the old one off, you can install the new one at home, but you must be prepared to heat the new bearing in hot oil at 325° F. maximum, the bearing resting on the pan bottom. When it is heated it becomes a slip fit on the bearing journal. Grasp the bearing with adequately protected tool or heavy rag, and with the bearing shield facing down, push it down onto the crankshaft. When it cools it will tighten perfectly. That's the way Briggs & Stratton recommends installing the new bearing, but if you take the crankshaft to the shop to have the old one pressed off, there is no reason why you couldn't have the new one pressed on.

In aluminum alloy engine models, the valve tappets are installed first, the crankshaft next, and then the cam gear. When installing the cam gear, turn the crankshaft and cam gear so that the timing marks on the gears align together. Model series 94000, 171700, 191700, 251700, and 252700 have a removable timing gear. When installing the timing gear, put the inner chamber toward the crankpin, thus assuring that the timing mark is visible.

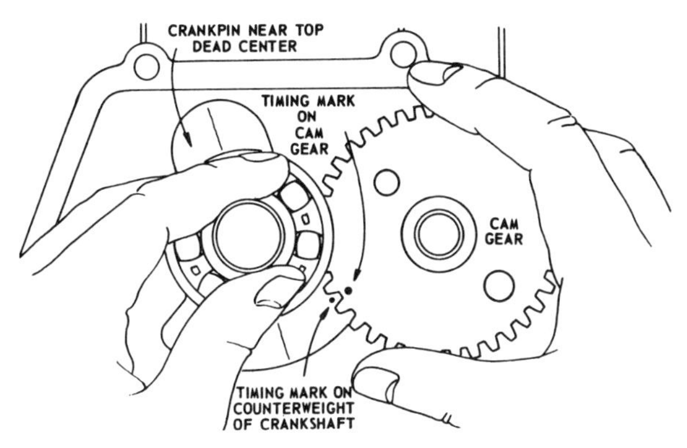

FIGURE 179. Timing mark alignment on ball bearing Briggs & Stratton engines.

MODEL SERIES	P.T.O. JOURNAL		MAGNETO JOURNAL		CRANKPIN JOURNAL	
ALUMINUM CYLINDER	**Inches**	**Millimeter**	**Inches**	**Millimeter**	**Inches**	**Millimeter**
6B, 60000	.873	22.17	.873	22.17	.870	22.10
8B, 80000*	.873	22.17	.873	22.17	.996	25.30
82000, 92000*, 94000, 110900*, 111200, 111900*	.873	22.17	.873	22.17	.996	25.30
100000, 130000	.998	25.35	.873	22.17	.996	25.30
140000, 170000	1.179	29.95	.997#	25.32#	1.090	27.69
190000	1.179	29.95	.997#	25.32#	1.122	28.50
220000, 250000	1.376	34.95	1.376	34.95	1.247	31.67
CAST IRON CYLINDER	**Inches**	**Millimeter**	**Inches**	**Millimeter**	**Inches**	**Millimeter**
5, 6, 8, N	.873	22.17	.873	22.17	.743	18.87
9	.983	24.97	.983	24.97	.873	22.17
14, 19, 190000	1.179	29.95	1.179	29.95	.996	25.30
200000	1.197	29.95	1.179	29.95	1.122	28.50
23, 230000†	1.376	34.95	1.376	34.95	1.184	30.07
240000	Ball	Ball	Ball	Ball	1.309	33.25
300000, 320000	Ball	Ball	Ball	Ball	1.309	33.25

*Auxiliary drive models P.T.O. Bearing Reject Size — 1.003 in. (25.48 mm).
#Synchro-Balance Magneto Bearing Reject Size — 1.179 in. (29.95 mm)
†Gear Reduction P.T.O. — 1.179 in. (29.95 mm)

FIGURE 180. Briggs & Stratton crankshaft reject table.

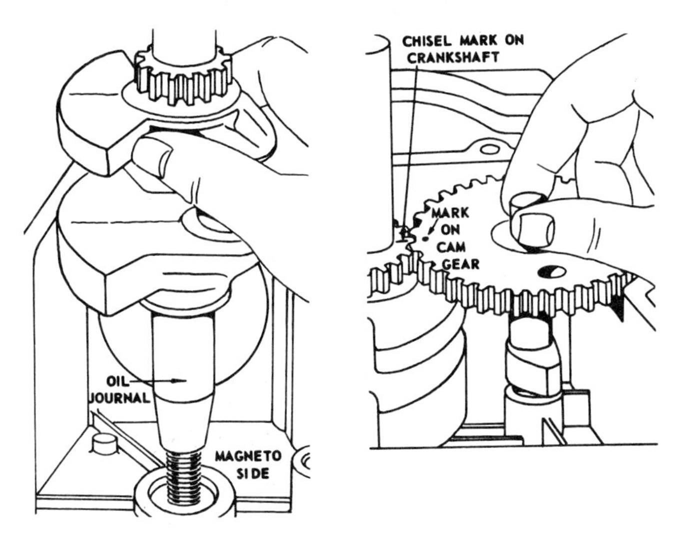

FIGURE 181. Aligning timing marks.

Timing marks are found in different places on different makes and models. Timing marks are a lot like trouble; if you go looking for them, you generally find them.

Crankshaft end play can be a problem. First, how do you know what to look for and where to look for it? The best way is to put something on the crankshaft at the point it enters (or leaves) the bottom cover or sump. Then use a feeler gauge as you push or pull the crankshaft. Measure the movement. You can use almost anything—from a chalk mark to a rubber band. The play should be between .002 and .008 inch. If it doesn't fall between those numbers, you have to use thrust washers to take up slack. They go between the crankcase cover or sump and the gear just above it. This is at the power takeoff end. That is the case, at least, in Briggs & Stratton aluminum engines with plain bearings, but aluminum engines with ball bearings add the thrust washer to the magneto end (opposite) of the crankshaft. On models 100900 and 130900 you must use a spring washer on the cam gear to take up excess end play. When putting back the crankcase cover it is important to

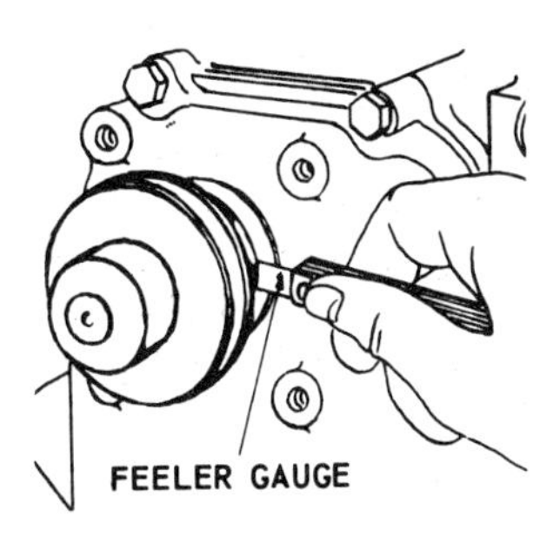

FIGURE 182. Measuring crankshaft end play with a feeler gauge.

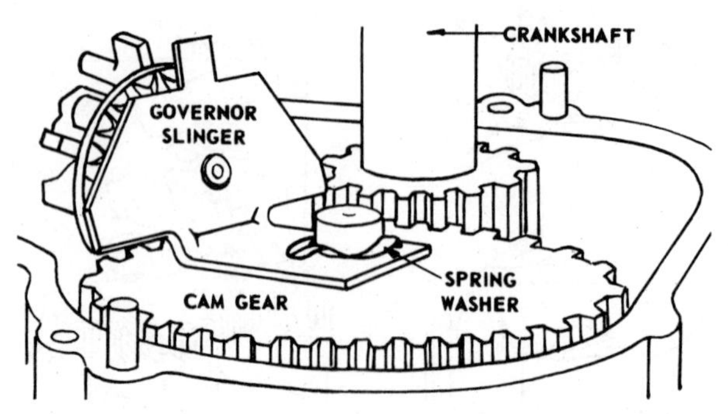

FIGURE 183. Spring washer shown here on cam gear to take up excess play.

protect the oil seal. The crankshaft may have sharp edges that might cut the seal, causing it to leak. You can wrap something over the crankshaft that will protect the seal through the rough spots—a piece of thin cardboard, for example. Also, put some oil or grease on the sealing edge of the oil seal.

Other small engines have similar end play adjustments and specifications.

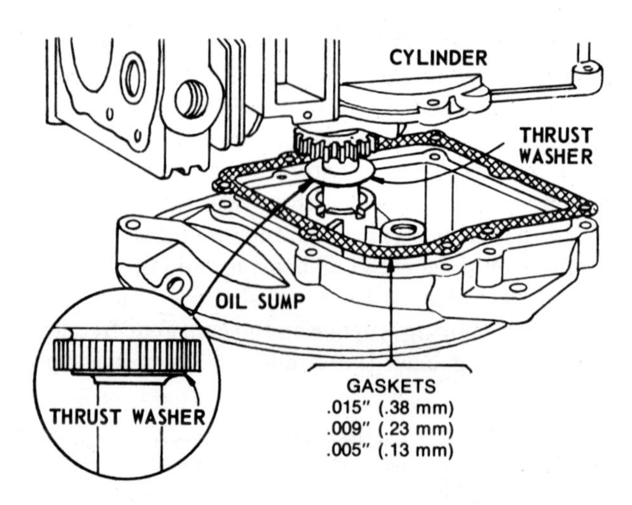

220624 THRUST WASHER for .875" (22.23 mm) DIA. CRKSFT.
220708 THRUST WASHER for 1" (25.40 mm) DIA. CRKSFT.
222949 THRUST WASHER for 1.181" (30.00 mm) DIA. CRKSFT.
222951 THRUST WASHER for 1.378" (35.00 mm) DIA. CRKSFT.

FIGURE 184. Correcting crankshaft end play requires gaskets and thrust washers in various Briggs & Stratton engines.

TWO-CYCLE ENGINES

A word about two-cycle engines should not conceal the fact that the key differences with engines discussed in previous sections are mostly based on the different valve systems. Some of these engines use reed valves rather than the valve system familiar in the other engines. Reed valves are flap or flutter valves that move according to crankcase pressure. Reduced pressure opens the valve, letting fuel air and oil mix into the crankcase. Increased crankcase pressure closes the valve, preventing escape of the mix. Carburetors on these engines may be found in queer places—at the lower crankcase or at the cylinder. Otherwise, most systems behave similarly to those we've discussed—fuel and carburetion, ignition, starting, etc.

One ingenious design, found in Tecumseh engines, is the two-cycle engine without valves. It substitutes ports that are opened and closed off by piston movement. Ports are small cylinder openings that permit gases to pass in and out of the combustion chamber.

The two-cycle engine uses a mixture of oil and gas. Oil enters with the fuel and clings to moving parts, lubricating bearing surfaces. It is vital that the right mix of oil and gas be used.

Reed valves build up deposits and sometimes need cleaning out, but otherwise they require little or no service.

Tecumseh recommends the following approach to troubleshooting, after you eliminate carburetion and ignition as possible sources of poor operation:

1. Crank engine slowly, checking for noise, binding, scraping, or other signs of improper performance. Look for damaged bearings, bent crankshaft or connecting rod.
2. Rock crankshaft back and forth to check for excessive play. That could indicate worn rod bearings or a worn piston pin.
3. Check seals at ends of crankcase for evidence of oil leaks, indicating a faulty seal. A leaking crankcase or crankcase seals will result in faulty fuel metering, erratic operation, and hard starting. For engines without lower crankcase seal, check bearing surface on lower half of crankcase.
4. On split crankcase models, check around entire crankcase for leaks where crankcase halves are joined. Leaks are usually indicated by oil deposits.
5. If applicable, remove the carburetor and check the reed plate assembly. Reeds should not be open more than .010 inch, and should not be warped, bent, chipped, or cracked.

The reference to split crankcase engines means primarily outboard motorboat engines, a special breed of engine that requires separate treatment. Lawn mower types do not have this complication of design.

Two-cycle engine care follows other engine maintenance. One difference arises—Tecumseh recommends that exhaust and engine ports be cleaned frequently to prevent carbon buildup. Also, to avoid damaging vibration, tighten all nuts and bolts regularly.

ELECTRIC MOTORS

The discussions of starter motors cover most problems that arise with outdoor equipment powered by electric motors. Such devices as electric compressors use motors that require the same maintenance and repair elaborated earlier. There are differences—power saws, compressors, drills, etc., use belts or gears to drive the equipment, rather than clutches as in starter motors. However, some electric motors that drive heavy equipment may have auxiliary starting devices on them such as capacitors. These devices, however, are usually not serviceable.

HEAVY EQUIPMENT

Snowblowers, tillers, riding equipment, and small tractors all use gasoline engines such as we've discussed. Many of them use auto-type transmissions, steering, and paraphernalia that require as much systematic treatment as we've given to engines.

The snowblower (-thrower, etc.) is the most familiar outdoor equipment north of the Mason-Dixon line. It consists of the kind of engine we've discussed, along with an auger that chews up the snow and spits it out through a chute or some other device. Some type of wheel drive, similar to the self-propelled lawn mower system, and simple clutches to engage the auger, turn the snowblower into a more complex machine than the lawn mower.

Troubleshooting these devices—apart from the engine—involves the auger and wheel drive. When the auger doesn't turn, first make sure it isn't stuck on a piece of tree or rock. Then check the auger drive lever to make sure it is engaged. Next look at the belts (not all makes and models use belts; some use gears). Check out the drive chain.

Is the snow falling all over you instead of being thrown out of the way? Check out the spout deflector for incorrect height and/or clogs. When the wheels don't turn properly, it can be caused by ice, but also the differential lock may not be engaged—it's the other control on the handles.

Another adjustment involves the distance between the scraper bar and the surface, which varies from the smooth surface of paving to stone surfaces. This is called the skid shoe adjustment.

These are all simple maintenance procedures and adjustments that any owner can be expected to perform. Cleaning, greasing, and lubrication are the most important aspects of maintenance. Grease fittings are provided on impeller bearings for regular greasing. Failure to do so, every 10 hours, could cause bearing failure and a costly, difficult job. Oiling of chains and surfaces with ordinary engine oil is another regular maintenance.

Snowblowers are long-suffering devices, if they don't get abused, but sticks and stones can indeed break their bones—auger, impeller, etc. It helps to avoid such hazards. However, the big problem in such equipment is always the engine.

Small tractors find many more uses nowadays than in the past. These are powered by the kinds of small engines we've discussed, but in the higher power ranges, say around 10 or higher horsepower.

They resemble small trucks in that there are steering, braking, transmission, frame, and suspension problems. One expects and finds similar types of maintenance and repair, depending on amount and severity of use.

INDEX